Climate Changes and Epidemiological Hotspots

Climate Changes and Epidemiological Hotspots

Debleena Bhattacharya
V K Singh

CRC Press
Taylor & Francis Group
Boca Raton London New York

CRC Press is an imprint of the
Taylor & Francis Group, an **informa** business

First edition published 2022
by CRC Press
6000 Broken Sound Parkway NW, Suite 300, Boca Raton, FL 33487-2742

and by CRC Press
2 Park Square, Milton Park, Abingdon, Oxon, OX14 4RN

© 2022 Debleena Bhattacharya and V K Singh

CRC Press is an imprint of Taylor & Francis Group, LLC

Reasonable efforts have been made to publish reliable data and information, but the author and publisher cannot assume responsibility for the validity of all materials or the consequences of their use. The authors and publishers have attempted to trace the copyright holders of all material reproduced in this publication and apologize to copyright holders if permission to publish in this form has not been obtained. If any copyright material has not been acknowledged please write and let us know so we may rectify in any future reprint.

Library of Congress Cataloging-in-Publication Data
A catalog record for this title has been requested

ISBN: 978-0-367-63768-2 (hbk)
ISBN: 978-0-367-63776-7 (pbk)
ISBN: 978-1-00-312062-9 (ebk)

DOI: 10.1201/9781003120629

Typeset in Times
by Newgen Publishing UK

Contents

Contents

Preface

"The world has enough for everyone's needs, but not everyone's greed,"
– Mahatma Gandhi

These lines always awakened me to think of the necessities of ours which has broadened the epidemiological hotspots of the environment. This book is the perspective of the authors; when we were writing this book the environmental damage had a new dimension in the form of a minuscule virus known as coronavirus that showed its prominence worldwide and restricted the originators in their dwellings. Time and again nature has shown us that the world that we are gifted is just a speck of white dot in the universe and our comfort zone has created the nuisance in the natural world.

The journey of the human race as defined by Darwin in *Descent* gave the thought that human thinking and their action in the aspect of the moral realm have evolutionary origin as like our physical nature.

This book is conceptualised with the vision of the present precarious times which had a profound impact on the environment and its dwellers.

The climate change that the world has witnessed during the lockdown for pandemic was altogether a different paradigm shift as the depletion of ozone hole was possible due to the lowering of the pollution coming out from the anthropogenic activities.

We often forget that this earth does not only constitute us, rather it is the house of many other species and every single organism has a definite role to play for the existence of life. Mankind has the virtue of being the most evolved species with an ability to make innovative changes. These exemplary qualities have nurtured the seed of greed in us that turned our virtues into vices and the consequences of this we are facing in the present times.

The journey of the book will help you to envision the underlying problems and the subtle changes from our ways to safeguard the environment.

I extend my gratitude to my co-author whose valuable inputs and guidance have always broadened my horizon.

The global risk perception to the metamorphosis will quench the thirst of knowledge and enlighten the horizon.

Acknowledgement

"A journey of a thousand miles begins with a single step".

– Lao-tzu

Sometimes the words cease to comprehend when you are surrounded with gratitude. When the journey started I was a bit apprehensive as it was a new dimension of my life that I am going to transverse. The word of motivation and kind support from Surgeon R Admiral V K Singh (my co-author who has already published two books on healthcare) has propelled the writer within me to embark on this journey. Thank you for being the constant support pillar and inspiration.

I would like to extend my gratitude to my parents who have always been my constant support and inspiration in every endeavour of mine. Every achievement of mine reflects the dedication, love and undulated support that both have fostered in me and gave me the wings to flutter in the limitless sky. Baba, you are my shining pole star in that infinite dark stretch, constantly guiding me and bestowing me with your blessings from above.

Thank you to my brother who has always been there for me and supported her little sister in every small step she has taken so far. I want to thank each and every member of my family for their constant support, blessings and love. Lastly, I want to thank the most precious member of the family, my dearest Pepsi, whose unconditional love has always uplifted my spirits.

In this present unprecedented situation, hurdles in the form of COVID-19 infection had shattered my hopes, but when you are bestowed with good people around you the journey smoothens. I want to thank my friends and all the people who have helped me during this journey.

Lastly, I owe my heartfelt gratitude to Dr Gagandeep Singh and my publisher CRC Press for giving me the opportunity and encouragement to accomplish this journey.

Author Biography.

DEBLEENA BHATTACHARYA

Debleena Bhattacharya holds a doctoral degree in Environmental Science from Indian Institute of Technology (Indian School of Mines) Dhanbad (India). Her research interest mainly focuses on two aspects: environmental biotechnology and wastewater treatment implemented with advanced technology.

At present she is a faculty member in the Department of Environmental Science and Engineering in Marwadi Education Foundation's Group of Institutions (MEFGI), Rajkot, Gujarat, India. Prior to joining MEFGI, she worked in SRISTI (Society for Research and Initiatives for Sustainable Technologies and Institutions as Project Coordinator, an NGO based in Ahmedabad, BIRAC (Biotechnology Industrial Research Assistance Council) for BIRAC-SRISTI PMU, a joint venture of Government of India and NGO, apart from being the laboratory co-ordinator of SRISTI lab.

SURGEON R ADMIRAL V K SINGH

Surgeon R Admiral V K Singh, VSM (Retd) is the Managing Director of InnovatioCuris (IC); Chairman of InnovatioCuris Foundation of Healthcare Excellence (ICFHE), a not-for-profit organisation; and adjunct professor at World Health Innovation Network, Odette School of Business, University of Windsor, Canada. He is also an honorary professor at Australian Institute of Health Innovation, Macquarie University, Australia, and a visiting professor at Indian Institute of Management, Indian Institute of Technology and others. He has authored two books, *Innovations in Healthcare Management: Cost Effective and Sustainable Solutions*, with national and international authors published by CRC Press in USA in 2015 and Special Indian edition in 2017, which owned prestigious Shingo publication award 2018, and *Planning and Designing Healthcare Facilities: Lean, Innovative and Evidence Based*, published in October 2017 by the same publisher. He launched a global magazine in healthcare innovation called *InnoHEALTH* in 2016. He holds an MBBS, a Master of Hospital Administration, a Diploma in National Board in Hospital and Health Care Management and an MPhil degree.

He has served as Deputy Chief Medical Officer of the United Nations and visited 42 countries. He has been awarded Vishisht Seva Medal (Distinguished Services Medal) for services to Zambia by the President of India and commendation of

Government of Zambia. He was an adjunct professor at Massachusetts Institute of Technology (MIT) Zaragoza and adjunct research professor at International Health Innovation Center, Ivey School of Business, University of Ontario, Canada. He was a consultant of National Disaster Management Authority and has formulated national guidelines for disaster on mass management of casualties; founder director of International Institute of Health Management and Research and president; is a fellow of Academy of Hospital Administration; chairman of Health Care Division of Quality Council of India; and has trained assessors and set standards for National Accreditation Board for Hospitals and Healthcare Providers (NABH) of India. He was the chairman of Healthcare Committee of Quality Council of India. He is a consultant in the Joint Commission International (JCI) accreditation system and an assessor of the European Foundation for Quality Management (EFQM). He has developed a health-specific EFQM model released in India by the Union Health Minister in December 2010. He had been the lead assessor of ISO and consultant and trainer in application of Lean principles in healthcare. He was the director of Healthcare Asia of Simpler dealing with Lean healthcare excellence, is a member of Telemedicine Society of India and Member of National Public Health Committee of Confederation of Indian Industry (CII), and planned and executed health/hospital projects by cutting down cost at planning, operations and transformation of health organisations in India and abroad by applying Lean innovations and evidence-based principles.

1 Introduction

WHY THIS BOOK?

The world is undergoing a troubled time when the very definition of safety has lost its eminence and we couldn't find a proper direction to resolve the predicament. Our ever-increasing carbon footprint has impacted the environment around us. We are in need of a change for the situation that we have created.

It is the propensity of human beings to develop a problem and thereafter search for its solution. The observable effects of climate change will show its prominence in the near future.

In the Paris Climate Accord held in 2015, countries were committed to reducing their carbon output and halt global warming to 2 or 1.5 degrees Celsius by the end of the century to reduce the impacts of climate change. According to the World Meteorological Organization (WMO) report, only 20% chances are there that in the next five years, it will be 1.5 degrees Celsius warmer than the pre-industrial levels and will continuously increase with time.

Though the recent pandemic has impacted the environment and climate, the nature of carbon dioxide (CO_2) to live long in the atmosphere will not be impacted by the recent reduction of CO_2 atmospheric concentration due to global lockdown of industries.

According to the data of the Intergovernmental Panel on Climate Change (IPCC), we need to find alternatives for our fossil fuel dependence by 2030 in order to prevent the global temperature rise to a threshold of 1.5 degrees Celsius, i.e. above pre-industrial levels. Premonitions are made by the experts that over-coming our threshold value will induce more heatwaves and hot summers, greater sea level rise, worst drought, heavy rainfall, wildfires, floods and eventually the food shortage of millions of people.

Due to the aggravation of the environment there is a risk of *Force Majeure* situation which can be construed by the series of epidemiological conundrums. This book specifically outlines the major environmental concerns that when monitored will conserve the sanctity of nature and prevent the harmful consequences foreseen in future.

DOI: 10.1201/9781003120629-1

WHAT IS CLIMATE CHANGE?

Climate change is a detrimental factor for the evolution of earth from its earlier version. According to the Paris Climate Accord 2015, climate has rendered a lot of changes in the surrounding. The Intergovernmental Panel on Climate Change (IPCC) suggested nine climatic tipping points that are connected both biologically and physically in complex ways, and if these happen the world will witness an irreversible change. The tipping points are the losses of the Amazon rainforest, Boreal forests, Permafrost, Arctic sea ice, Coral Reef, Atlantic circulation, Greenland ice sheets, Wilkes Basin Antarctica and west Antarctic ice sheets. The risk of natural calamities increases with the change in climate and the subsequent changes has increased the frequency of environmental concerns such as heatwaves, drought, floods and natural disasters like hurricanes, etc. Climate change will also have an impact on crop yield and increase in vector-borne diseases (Upadhyay et al., 2013).

WHAT ARE THE EPIDEMIOLOGICAL HOTSPOTS?

The environment is greatly hampered by constant changes occurring in the surroundings; therefore, in order to safeguard our health, we identify the hotspots.

Climate change has affected the health of people in various ways, i.e. through increased natural calamities where there is a frequency of heatwaves, floods, droughts and landslides. There are many diseases of significant public health disruptions, such as malaria and dengue, which are highly prone to change in climate (Patz and Kovats, 2002).

The burgeoning waste disposal before and presently amid the recent pandemic also acts as an ignition to the change in the environment. The waste in the form of both liquid and solid contributes to the dilapidation of the surroundings. The obnoxious gases and the harmful residuals emitted when these waste and wastewater are unattended for treatment lead to a toxic atmosphere. The accumulated waste is a nuisance to the living beings who dwells nearby it. The spread of contagion diseases is more rapid when they channelise through waste.

The increase in pollution due to expansion of cities and removal of vegetation will have detrimental impact on the environment because of global warming and this is very well foreseen by the impact on ozone layer. Ozone therefore has severe consequences for people with asthma (McConnell et al., 2002) as they are sensitive to these allergens and it promotes the development of asthma in children in their initial growing years (Koren and Bromberg, 1995; Koren and Utell, 1997).

This book enlightens about the various hotspots that remain unattended now will eventually find their significance in the long run.

Before the onset of the pandemic, the world has suffered the fury of nature for the last 49 years where 50% disasters occurred from weather, climate and water hazards, death toll were reported at 45% and 74% economic losses were incurred (World Meteorological Organization, 2021). These events remind us about our fragile inheritance and the need to conserve our environment.

Risk from flooding by the coastal storm is predicted to escalate from the present scenario of 75 million to 200 million of midrange climate changes, and the rise in the sea level is envisioned to reach to 40 cm by the 2080s (McCarthy et al., 2001).

The increase in seas might result in salination of coastal freshwater aquifers and also hamper the storm water drainage and sewage disposal therefore creating a situation where drinking water availability will be a concern (Patz, 2001).

The contaminated water will pave the path for communicable diseases (e.g. cholera or hepatitis A) and occurrence of disease outbreaks with flood waters (e.g. leptospirosis). Studies reported the epidemic of leptospirosis, a bacterial zoonosis that occurred after floods in Nicaragua and Brazil (Trevejo et al., 1998; Centers for Disease Control and Prevention, 1995; Ko et al., 1999).

Another study showed that in Bangladesh in 1988, watery diarrhoea predominant in the population after floods was the most common factor for death of all age groups under 45, followed by respiratory infection (Siddique et al., 1991).

Frugal innovation in all the sectors will help us to accomplish the 17 sustainable development goals.

With the paradigm shift in climate, the food insecurity, risk of drought and lack of resources to import food will eventually increase the malnourishment in the population.

In a few years as projected (i.e. 2025), there will be an acceleration of 5 billion people who inhabit water-stressed countries.

Though technology has been a boon to the agriculture industry with improved crop varieties and irrigation system, the agricultural productivity backbone still lies in the climatic conditions. The change in environment will show its repercussions in the yield of crops (McMichael et al., 2001).

The other hotspots are the regions that experience weather extremes due to El-Nino weather patterns (Peru and Ecuador for floods, Southern Africa; Indonesia and Malaysia for droughts; Malaria in Punjab, India; and Cholera in Bangladesh) (Patz and Kovats, 2002).

This book will give a brief idea about the way the problem is perceived and the innovative way to resolve the same. The impact of climate change on our environment and surroundings is profound and our negligence eventually will pave the grave for us.

The recent pandemic gave us the outlook that even the most well-equipped and powerful nation has its own limitations and had to face the consequences of ignorance. The epidemiological hotspots when not addressed for a longer time will have a severe impact on the climate.

REFERENCES

1.   Upadhyay, N., Sun, Q., Allen, J.O., Westerhoff, P. and Herckes, P., 2013. Characterization of aerosol emissions from wastewater aeration basins. *J Air Waste Manag Assoc*, *63*(1), 20–26.
2.   Patz, J.A. and Kovats, R.S., 2002. Hotspots in climate change and human health. *BMJ*, *325*(7372), 1094–1098.

3. Koren, H.S. and Bromberg, P.A., 1995. Respiratory responses of asthmatics to ozone. *Int Arch Allergy Immunol, 107*, 2368.

4. Koren, H.S. and Utell, M.J., 1997. Asthma and the environment. *Environ Health Perspect,; 105*, 5347.

5. McConnell, R., Berhane, K., Gilliland, F., London, S., Islam, T., Gauderman, W., et al., 2002. Asthma in exercising children exposed to ozone: a cohort study. *Lancet, 359*, 38691.

6. McCarthy, J., Canziani, O., Leary, N., Kokken, D. and White, K., 2001. *Climate change 2001: impacts, adaptation, and vulnerability.* New York: Cambridge University Press. (UN Intergovernmental Panel for Climate Change. Third assessment report.)

7. Patz, J., 2001. Public health risk assessment linked to climatic and ecological change. *Hum Ecolog Risk Assess, 7*, 131727.

8. Trevejo, R.T., RigauPerez, J.G., Ashford, D.A., McClure, E.M., JarquinGonzalez, C., Amador, J.J., et al., 1998. Epidemic leptospirosis associated with pulmonary hemorrhage—Nicaragua, 1995. *J Infect Dis, 178*, 145763.

9. Centers for Disease Control and Prevention, 1995. Outbreak of acute febrile illness and pulmonary hemorrhage—Nicaragua, 1995. *Morb Mortal Wkly Rep MMWR; 44*, 8413.

10. Ko, A.I., Galvao Reis, M., Ribeiro Dourado, C.M., Johnson, W.D. Jr and Riley, L.W., 1999. Urban epidemic of severe leptospirosis in Brazil. Salvador Leptospirosis Study Group. *Lancet, 354*, 8205.

11. Siddique A, Baqui A, Eusof A and Zaman K, 1991. 1988 floods in Bangladesh: pattern of illness and cause of death. *J Diarrhoeal Dis Res, 9*, 3104.

12. McMichael A, Githeko A, Akhtar R, Carcavallo R, Gubler D, Haines A, et al., 2001. Human health. In: McCarthy J, Canziani O, Leary N, Dokken D, White K, eds. *Climate change 2001: impacts, adaptation, and vulnerability.* New York: Cambridge University Press.

13. https://unfccc.int/sites/default/files/english_paris_agreement.pdf

14. https://public.wmo.int/en/media/press-release/weather-related-disasters-increase-over-past-50-years-causing-more-damage-fewer

2 Global Risk Perception

THE SUBTLE GLOBAL RISK

Whenever we think of making a product we never think about the amount of waste it will generate to make the same. The waste might be in various forms, textures, locations, states and nature. The concept of "cradle to grave" when followed gives an insightful journey and helps us to focus on the global risk.

This chapter transcends you to the journey of understanding the global risk that surrounds the environment.

According to the Global Risk Report 2020, these risks can be further categorised into the following:

1. Extreme weather due to geopolitical instability
2. Failure of climate change mitigation and adaptation
3. Climate response shortcomings
4. Massive biodiversity loss
5. Major natural disasters: earthquakes, tsunamis, volcanic eruptions, and geomagnetic storms
6. Staggering health system
7. Technological governance deficits
8. Human intervention for the environmental damage and disaster
   - Oil spills
   - Effluent
   - Radioactive wastes

The geopolitical risk has increased a lot of global tensions between nations. The burning of the Amazon forest had a huge impact on the climate and the environment as a whole. The geopolitical propaganda connects Brazil to China, China to the United States (US) and Brazil, as a member of the Mercosur to the European Union (EU). According to the United Nations (UN), "global food production is responsible for 21-37% of 'anthropogenic emissions' " (www.opendemocracy.net).

DOI: 10.1201/9781003120629-2

The UN's International Panel on Climate Change (IPCC) report *"Climate Change and Land"* emphasised the urgency of addressing the issues of land use, food production and consumption as "food production is the largest cause of global environmental change". It accounts for up to 37% of total global greenhouse gases (GHGs), and two and a half times more than that caused by total global transport (www.opendemocracy.net).

The demand for food and the global responsibility of being a self-sufficient nation with a proper food security have paved the way for excessive use of fertilisers in the agricultural land of China. The ecological catastrophe was witnessed by China and UN Food and Agriculture Organization, where excessive use of fertilisers has increased the global production of nitrogen gas. As nitrogen is a powerful greenhouse gas, it invariably affects the climate. The demand for soya bean production in Brazil to sustain meat production in China has led to the clearance of the Amazon forest. The impact of the forest fire of Amazon has led to the spread of pollutants in the form of particulate matter and toxic gases such as carbon monoxide, nitrogen oxides and non-methane organic compounds.

The upcoming years will witness significant changes in the climate and as a result the change in weather conditions. The global emission needs to be reduced to safeguard our planet from the harmful consequences and its effects by declining it to 7.6% annually between 2020 and 2030. The holistic approach towards transition will help us for long-term goals. We can understand the same with an approach that uses low carbon technologies for various metals like copper, cobalt and manganese that can be mined from the seabed, but the impacts of deep sea can disrupt the benefits. Large areas of land converted to monocultures had an ulterior motive for food production and nature. Geopolitical relationships have changed the focal point in the trade in fossil fuel as it has become economically less important.

INNOVATIVE INTERVENTIONS

Beneath the dark clouds hides the shining sun during the time of eclipse; likewise the technological breakthrough is a ray of hope hidden under the risks. An example is where a start-up has come up with a novel idea to develop a way to harness artificial intelligence and mirrors reflecting the sun to create the extreme heat required for industrial processes – a potential game changer for the source of around 10% of global emissions each year.

The amount of clean energy is increasing and it is also getting cheaper alongside creating job opportunities. The onshore and photovoltaic solar power cost has subsided over the decade by 70% and 90%, respectively. In many nations the cost of installation of new coal-based power plant is rather costly than a wind and solar power station. Shifting to renewable energy sources could grow the world economy and the potential industry for the coming decade is through meat production and agriculture.

Failure of climate change and mitigation will pose a problem for the aggravation of global risks. The change in climate is dependent on natural and man-made causes where the anthropogenic criteria had a deep impact on the climatic adversities.

The sixth assessment report of IPCC provided the constraining factors with potential application and mitigation as described in Table 2.1. Adaptation and

TABLE 2.1
Summary of the Constraining Factors with Potential Application and Mitigation

Constraining Factor	Potential Implications for Adaptation	Potential Implications for Mitigation
Adverse externalities of population growth and urbanisation	Increase exposure of human populations to climate variability and change as well as demands for, and pressures on, natural resources and ecosystem services	Drive economic growth, energy demand and energy consumption, resulting in increases in greenhouse gas (GHG) emissions
Deficits of knowledge, education and human capital	Reduce national, institutional and individual perceptions of the risks posed by climate change as well as the costs and benefits of different adaptation options	Reduce national, institutional and individual risk perceptions, willingness to change behavioural patterns and practices and to adopt social and technological innovations to reduce emissions
Divergences in social and cultural attitudes, values and behaviours	Reduce societal consensus regarding climate risk and therefore demand for specific adaptation policies and measures	Influence emission patterns, societal perceptions of the utility of mitigation policies and technologies, and willingness to pursue sustainable behaviours and technologies
Challenges in governance and institutional arrangements	Reduce the ability to coordinate adaptation policies and measures and to deliver capacity to actors to plan and implement adaptation	Undermine policies, incentives and cooperation regarding the development of mitigation policies and the implementation of efficient, carbon-neutral and renewable energy technologies
Lack of access to national and international climate finance	Reduces the scale of investment in adaptation policies and measures and therefore their effectiveness	Reduces the capacity of developed and, particularly, developing nations to pursue policies and technologies that reduce emissions.

(continued)

TABLE 2.1 (Continued)
Summary of the Constraining Factors with Potential Application and Mitigation

Constraining Factor	Potential Implications for Adaptation	Potential Implications for Mitigation
Inadequate technology	Reduces the range of available adaptation options as well as their effectiveness in reducing or avoiding risk from increasing rates or magnitudes of climate change	Slows the rate at which society can reduce the carbon intensity of energy services and transition towards low-carbon and carbon-neutral technologies
Insufficient quality and/or quantity of natural resources	Reduce the coping range of actors, vulnerability to non-climatic factors and potential competition for resources that enhances vulnerability	Reduce the long-term sustainability of different energy technologies
Adaptation and development deficits	Increase vulnerability to current climate variability as well as future climate change	Reduce mitigative capacity and undermine international cooperative efforts on climate owing to a contentious legacy of cooperation on development
Inequality	Places the impacts of climate change and the burden of adaptation disproportionately on the most vulnerable and/or transfers them to future generations	Constrains the ability for developing nations with low income levels, or different communities or sectors within nations, to contribute to GHG mitigation

Source: Sixth assessment report of IPCC (2014).

mitigation approaches face the constraints of inertia of regional and global trends with the GHG emission, development of the economy, use of resources patterns implemented for infrastructure and settlement along with the institutional technology and characteristics. India is already facing the scorching heat of summer and an exponential growth in coronavirus disease 2019 (COVID-19) cases. Many places of India have crossed 40 degrees temperature, among them Odisha is one that has shown an increase in temperature in early February 2021.

North Indian states are surpassing all the records pertaining to rising heat and higher than normal temperatures. The year 2021 is the La Niña year where Pacific currents bring cooler temperatures but global warming has diminished the effects.

The increase in temperature will have a disastrous effect on water security. There will be greater evaporation from all water bodies. The more the evaporation will be it will increase the scarcity of water.

A study reported in 2018 promulgated the idea that our ancient scriptures have the solution to these problems if we go through the old historical text. The ancient Mauryan dynasty had brought water in the dry lands of Magadh (Bihar) reviving the dilapidated network of pynes and ahars (reported in www.hindustantimes.com).

The pynes are channels carrying water from rivers and ahars are the low-lying fields with embankments acting as water reservoirs. This combined irrigation and water conservation system dates back to the Mauryan era that flourished in Magadh 2,000 years ago. The results were miraculous and instantaneous as about 150 villages along the Jamune-Dasainpine and about 250 villages along the Barkhi canal had the opportunity to irrigate their fields for the *kharif* and *rabi* (monsoon and winter) crops along with growing vegetables, pulses and oilseeds (reported in www.hindustantimes.com).

India has taken least concern for its groundwater management system as lots of irrigation bureaucracies have given more importance for canal and surface water implications. However, with the unpredictable change in climate there has to be an alternative in approach for ground water conservation along with a restriction of water loss from tanks, ponds, canals and other surface water sources. Loss of water due to evaporation has increased with time as there is an immense change in temperatures with the climatic change.

The soaring temperature will also be a factor for increased moisture loss from the soil, thereby making it arid and increasing the need for irrigation. According to the UN report, the breadbasket of the world (USA) is at risk due to changes in climatic conditions.

As per the report published by the UN (www.un.org), from 2010 to 2020, it was estimated that up to 44% of all the world's cultivated systems are in the drylands. Plant species endemic to the drylands make up 30% of the plants under cultivation today.

It was estimated that under the climatic change scenario, half of the world's population in 2030 will be living in areas of high water stress.

In a country like India where the majority of irrigation depends on rain water, the rain-fed regions will face the harsh consequences of water scarcity as it will intensify land degradation as well as dust bowl formation, leading to deprivation of food sources. This also gives the introspection that water management should be simultaneously taken forward along with vegetation planning to help in sustainability of the soil to hold water even during scorching heat conditions.

The change in climate will also give rise to the increase in forest fires, and in the recent past we have witnessed some major incidents that have devastating effects on the flora and fauna, as well as the environment as a whole. One such incident is the Australian forest fire which had a deep impact on the stratosphere level as it has altered the large-scale wind patterns with more than 10 miles overhead (Washington Post, 2020).

As mentioned in IPCC 2014a, the effects of global climate change that are currently experienced, like climate change-related extreme weather events, wildfires and water level rise, are only mild harbingers of projected future climate damages.

THE UNFORESEEN WORLD

The 2020 coronavirus pandemic may cause an in-depth understanding of the ties that bind us all on a worldwide scale and that will help us to overcome with the most important public health threat of the century, i.e. the climate crisis.

The COVID-19 pandemic taught us a crucial lesson by resonating its similarities with global climate change where well-resourced, equitable health systems with a robust and supported health workforce are essential to guard us from health security threats, including global climate change. The stringent measures that have strained many national health systems over the past decade will need to be reversed if economies and societies are to be resilient and prosperous in an age of change. According to the Global Climate Index 2021 data, vulnerable countries will be most affected with various types of risk, such as climatic, geophysical, economic and health-related risks.

For example, the people of Haiti are much more adept in handling and recovering from the lasting effects of Hurricane Matthew 2016 – which was exacerbated by global climate change – and it was made possible because they had had a resilient and well-resourced health system in place to support them. Similarly, many Iranian lives could have been saved at the primary stages of the COVID-19 outbreak in the country if its beleaguered healthcare system had been better prepared for what was to return.

The current unprecedented situation due to the pandemic further illustrates the role of inequality that plays a major barrier in ensuring the health and well-being of people, apart from the social and economic inequality that paves the way for unequal access to healthcare systems. For example, the health threat of the novel coronavirus is, on average, greater for cities and other people exposed to higher levels of pollution, which are most frequently people living in poorer areas. The same is true for the health impacts of global climate change, with one among its major causes, the burning of fossil fuels which also adds pollution to the air and disproportionately impacts the health of those in poverty.

The World Health Organization (WHO) estimates that by reducing the environmental and social risk factors that people are more vulnerable to, nearly 1/4 of the worldwide health burden (measured as loss from sickness, death and financial costs) could be prevented. Creating healthy environments for healthier populations and promoting Universal Health Coverage (UHC) are two of the foremost effective ways through which we can reduce the long-term health impacts – and increase our resilience and adaptive capacity to – both the coronavirus pandemic and global climate change.

The global health crisis has changed our perceptions to safeguard ourselves as well as those residing around us. The alteration in our outlook has benefitted us by inculcating the need for collective action and effective ways to manage the risks. Though climate change prophecies a slower, long-term health effect in near future, the sustainable change in approach to deal with the harmful impact will prevent irreversible damage. The crisis often leads to realisation and gives us the introspection for developing an opportune situation for the health and safety of mankind.

Biodiversity is the key to human existence on Earth, and the proof is unequivocal – it is being annihilated by us at a rate phenomenal ever. Since the advent of industrial revolution, human exercises have progressively annihilated and corrupted timber-lands, prairies, wetlands and other significant environments, compromising human prosperity. Of the Earth's sans ice land surface, 75% has effectively been altogether adjusted, the vast majority of the oceans have lost its pristine elegance to pollution and over 85% of the region of wetlands has been lost.

The direct driver of biodiversity loss in the last several decades has been the use of land, primarily the manifestation of native habitats (forests, mangroves and grasslands) to agricultural systems, and overfishing has led to the depletion of water bodies.

The important challenge lies in preventing the unethical practices for agriculture that lead to unsustainable situations. To quench the dire necessity to produce affordable and nutritious food, we need to conserve and protect biodiversity.

Though till date climate change has not been the most important cause of the loss of biodiversity, it is still projected to become one in the coming years. The loss of biodiversity adversely affects the climate and deforestation has increased the CO_2 level in the atmosphere as reported by the Intergovernmental Panel on Climate Change (IPCC) (www.ipcc.ch, 2020).

THE PROJECTED DETRIMENTAL IMPACT

The IPCC in its Fifth Assessment Report has emphasised the present and the projected impacts of anthropogenic activity on climate changes and extreme weather events.

The unprecedented global climate has exacerbated the climate hazards and amplified the risk of extreme weather conditions. The increase of air and water temperatures had led to the rise of sea levels, supercharged storms and a higher wind speeds, severe and prolonged droughts and a concurrent season of wildfire, more precipitation and flooding. As per the estimated data reported by the United Nations Environment Programme, adapting to climate change and coping with damages will cost developing nations around $140–300 billion per year by 2030. The world in the past decade has faced cyclones like Idai and Kenneth. In 2019, Cyclone Idai showed its prominence when more than 1,000 people across Zimbabwe, Malawi and Mozambique in Southern Africa faced death, and caused devastation to millions as they were left destitute without food or basic services.

The year 2020 began with the devastating bushfire season in Australia and a record hottest year was projected that has changed the fertile soil into an arid region. The start of 2020 found Australia in the midst of its worst-ever bushfire season – following on from its hottest year on record which had left soil and fuels exceptionally dry. The fires have burnt through more than 10 million hectares, killed at least 28 people, razed entire communities to the ground, taken the homes of thousands of families and left millions of people affected by a hazardous smoke haze. More than a billion native animals have been killed, and some species and ecosystems may never recover.

The impact of the Australian forest fire was seen by the formation of pyro-cumulo-nimbus (pyroCb) clouds that caused stratospheric perturbations, which has associated larger magnitude than the previous benchmarks of maximum pyroCb activity and had approached the impact of moderate volcanic eruption (Khaykin et al., 2018; Peterson et al., 2018). The volcanic eruptions inject the ash and sulphur which is further oxidised and condenses to form sub-micron-sized aerosol droplets into the stratospheric levels. With the PyroCb, intense fire-driven convection lifts combustion merchandise in vola-tilised type furthermore as stuff together with organic and black carbon, smoke aero-sols and condensed water. The star heating of the extremely assimilatory black carbon propels the smoke-laden air parcels upwards (Khaykin et al., 2018), which, combined with horizontal transport (Bourassa et al., 2019; Kloss et al., 2019), results in a lot of economical meridional dispersion of those aerosols and prolongs their stratospheric residence time (Yu et al., 2019).

According to a study, severe droughts in East Africa have left 15 million people in Ethiopia, Kenya and Somalia in need of aid, where only 35 percent is funded (www.oxfam.org/en/5-natural-disasters-beg-climate-action). Africa has witnessed horrific drought conditions in the past few decades that have harmed the livelihood and crops leading to famine conditions (www.oxfam.org).

The impact of floods has increased due to rising sea level as sea surface tem-perature has led to deadly floods and landslides in India, Nepal and Bangladesh, and these have forced 12 million people from their homes. The monsoon has inten-sified in the last few years as the eastern and northeastern region of India faced floods that caused huge loss to people and their livelihoods.

For six consecutive years the El Niño period, laden with the climate crisis, has impacted Central America's dry corridor. Guatemala, Honduras, El Salvador and Nicaragua witnessed their three months dry season surpass the period. Since most crops have failed, it has left 3.5 million people, many of whom rely on farming for both food and livelihood, in need of humanitarian assistance, and 2.5 million people food insecure.

The year that has gone by gave us a lot of insights apart from the climatic situ-ations. There is a revision on the benchmark of developed countries as the pan-demic showed the true strength of a nation lies in its healthcare system and if that is staggered then that nation is bereft of backbone and needs support to sustain itself.

The increasing death due to unavailability of proper medical facilities is the biggest loss of mankind which thrives on the notion of technological edge. There should be smart people around to run the advanced gadgets. The dilapidated situ-ation of healthcare is a big concern and the way the healthcare waste is handled is also adding to it.

EPIDEMIOLOGICAL INTERVENTION

The epidemiological statistics show that climate change is a deep-seated framework for causality and requires innovative approaches to eliminate it.

Environmental epidemiologists have studied the health effects of climate change and the major hotspots lie in the toxicological change in the environment pertaining to temperature or seasonal changes along with change in cellular responses, plant responses, phenological changes and non-human biological impact of climate-related changes. We are on the verge where we have exhausted our fossil fuel; excessive use of large establishments that consume massive amounts of electrical energy and generation of GHGs in the present scenario have also added to the above concern.

Habitat loss has forced animals to migrate and has potentially made it to come in contact with other animals or people and share the germs. Large livestock farms serve as a source for spreading of infections from animals to people. The decrease in demand for animal meat and achieving more sustainable animal husbandry could therefore reduce the emerging infectious disease risk and in future it could also lower greenhouse gas emissions.

REFERENCES

1. www.opendemocracy.net/en/oureconomy/what-amazon-fires-tell-us-about-geopolitics-climate-emergency/.
2. https://reports.weforum.org/global-risks-report-2020/.
3. www.hindustantimes.com/india-news/ancient-mauryan-technology-brings-water-hope-to-dry-magadh-in-bihar/story-aMtlAukgnrYEFR4vjB1DrL.html.
4. www.un.org/en/events/desertification_decade/whynow.shtml.
5. www.washingtonpost.com/weather/2020/06/22/australias-fires-blew-smoke-19-miles-into-sky-similar-predicted-nuclear-blast/.
6. IPCC., 2014. Climate change 2014: impacts, adaptation, and vulnerability. Part A: global and sectoral aspects. In: Field C.B., Barros V.R., Dokken D.J., Mach K.J., Mastrandrea M.D., Bilir T.E., Chatterjee M., Ebi K.L., Estrada Y.O., Genova R.C., Girma B., Kissel E.S., Levy A.N., MacCracken S., Mastrandrea P.R. and White L.L. (eds.), *Contribution of working group II to the fifth assessment report of the intergovernmental panel on climate change.*, Cambridge: Cambridge University Press.
7. www.weforum.org/agenda/2020/04/climate-change-coronavirus-linked/.
8. https://germanwatch.org/sites/default/files/Global%20Climate%20Risk%20Index%202021_1.pdf.
9. www.ipcc.ch/site/assets/uploads/sites/4/2020/02/SPM_Updated-Jan20.pdf.
10. Khaykin, S.M., Godin-Beekmann, S., Hauchecorne, A., Pelon, J., Ravetta, F. and Keckhut, P., 2018. Stratospheric smoke with unprecedentedly high backscatter observed by lidars above Southern France. *Geophys Res Lett, 45*, 1639–1646.
11. Peterson, D.A. Campbell, J.R., Hyer, E.J., Fromm, M.D., Kablick III, G.P., Cossuth, J.H. et al., 2018. Wildfire-driven thunderstorms cause a volcano-like stratospheric injection of smoke. *NPJ Clim Atmospheric Sci, 1*, 30.

12. Bourassa, A.E., Rieger, L.A., Zawada, D.J., Khaykin, S., Thomason, L.W. and Degenstein, D.A., 2019. Satellite limb observations of unprecedented forest fire aerosol in the stratosphere. *J Geophys Res Atmospheres, 124*, 9510–9519.

13. Kloss, C. Berthet, G., Sellitto, P., Ploeger, F., Bucci, S., Khaykin, S., et al., 2019. Transport of the 2017 Canadian wildfire plume to the tropics via the Asian monsoon circulation. *Atmos Chem Phys, 19*, 13547–13567.

14. Yu, P., Toon, O.B., Bardeen, C.G., Zhu, Y., Rosenlof, K.H., Portmann, R.W. et al., 2019. Black carbon lofts wildfire smoke high into the stratosphere to form a persistent plume. *Science, 365*, 587–590.

15. www.oxfam.org/en/5-natural-disasters-beg-climate-action.

3 Attributable Burden of Waste

WASTE: REGRETS OR RESOURCE

The focus of this chapter is on the contemporary wastes that are less discussed but equally relevant for environmental disruption and epidemiological cause. The word "waste" has varied forms that may be solid or liquid. The definition got more importance in advancement of technology in the form of electronic waste (e-waste).

When the famous chemist Leo Hendrik Baekeland discovered plastic it was a boon for mankind but little did he know that this will one day become the biggest nuisance that world will ever face. Till today the massacre that it has done to the environment is insurmountable. The advancement of technology gave us the opportunity to use computers, smart devices, etc.

The endless list of advanced methods to help in making the life easier for an individual has invariably paved the way for various health risks. According to scientific reports published in 2014, the generation of e-waste was nearly 20–50 million tons and as per recent data in 2019 it has crossed to 53.6 million metric tons (Mt) (Baldé et al., 2017). As per the study of Basel Action Network (Hussain and Mumtaz, 2014), 500 million computers contain 287 billion kg plastics, 716.7 million kg lead and 286,700 kg mercury. Apart from these the e-waste carries over 50 elements from the periodic table. The lethal heavy metal component (that consists of cadmium, mercury, copper, nickel, lead, barium, hexavalent chromium and beryllium), plastics, phosphor as well as brominated flame retardants are also significant. The persistent, mobile and bio-accumulative toxins that remain in the environment with an altered form lead to harmful effect on the body by being carcinogenic, mutagenic and teratogenic. The ensuing hazardous waste has impacted the physical, biological and socio-economic environments, giving rise to fatal diseases like cancer, reproductive disorder, neural damages, asthmatic bronchitis, endocrine disruption and brain retardation. The people who are living in slums are the most vulnerable as many of them work as ragpickers (Zolnikov et al., 2021).

DOI: 10.1201/9781003120629-3

HEAVY METAL ACCUMULATION

The recent accumulated data suggest that the generation of e-waste is more from developed countries and they are being transferred to developing countries for further processing (Rautela et al., 2021). China, India and Ghana are the major developing countries where e-waste is processed and the specific hotspot areas where they go for informal recycling are Guiyu, Taizhou and Longtang in China (Li and Achal, 2020; Lu et al., 2015); Delhi, Mumbai, Kolkata and Moradabad in India (Joon et al., 2017; Kumar et al., 2018); and Agbogbloshie and Accra in Ghana (Amoyaw-Osei et al., 2011). These are further elucidated in Table 3.1.

The pandemic has compounded the use of electronic devices as lockdown on business, education sector and other sectors has imposed a gap which was mended by the use of information, communication and technological (ICT) devices. The increased use of electronic devices (Liu et al.,2011) will lead to a more serious problem in the future and post-COVID-19 era. Already there are health hazards due to burgeoning e-waste and in the present scenario it will increase more. According to the United Nations Environment Programme (UNEP) and Baldé et al. (2017), e-waste will increase to 52 million Mt by 2021, and 120 million tons yearly by 2050, of which only 20% would be recycled. The e-waste is more like a dormant volcano

TABLE 3.1
The Concentration of Heavy Metals into the Environment Increases due to Informal Recycling of E-waste (Rautela et al., 2021)

Sr. no	Pollutants	Concentration Unit (ng/m³)	Treatment	Location	References
1	Benzophenones (BZP)	1.022	Mechanical treatment	Shijiao, Qingyuan city (China)	Li and Achal (2020)
2	Amine co-initiators (ACI)	0.169	Mechanical treatment	Shijiao, Qingyuan city (China)	Li and Achal (2020)
3	Thioxanthones (TX)	0.00459	Mechanical treatment	Shijiao, Qingyuan city (China)	Li and Achal (2020)
4	Cadmium (Cd)	1.2	Leaching processes	Guiyu (China)	Li et al. (2011)
5	Chromium (Cr)	54	Crude recycling	Bangalore (India)	Ha et al., (2009)
6	Cadmium (Cd)	4	Mechanical treatment and open burning	Agbogbloshie and Ashaiman, Accra (Ghana)	Ackah (2019)
7	Arsenic (As)	6	Mechanical treatment and open burning	Agbogbloshie and Ashaiman, Accra (Ghana)	Ackah (2019)

that will become active in the coming years and will lead to a very harmful impact in future. According to Manish and Chakraborty (2019), the amount of emissions, waste generation and leaching of the hazardous e-waste has impacted the climate as open dumping of e-wastes has generated several environmental issues like groundwater contamination, soil pollution and human health problems.

The National Aeronautics and Space Administration (NASA) followed the glacial mass thickness in Alaska from 2003 to 2010. In that seven-year period, 32 gigatons (322 million tons) of ice melted. Several factors are prompting an Earth-wide temperature boost. Contaminations in the water and air are driving reasons. At the point when e-squander isn't reused as expected, it impacts a worldwide temperature alteration.

As indicated by the United States Environmental Protection Agency (U.S. EPA), each 1 million cells that are reused recuperate 35,000 pounds of copper, 772 pounds of silver, 75 pounds of gold and 33 pounds of palladium. You'll dispense with the harm to the earth and the energy it takes to mine those metals or substance elements. Palladium is a cancer-causing agent that has been found in dust particles and air. Studies have found that it can harm the liver and kidneys. It has likewise been found to affect bone marrow. Mercury is additionally found in hardware and is connected to issues with stomach-related, resistant and sensory system.

WASTE GENERATION: BOON OR BANE

Apart from the e-waste (UNEP, 2019), the composition of these wastes has a huge amount of plastics and the major factor for changing of our climate and the generation of epidemiological hotspots has been attributed to plastic. Discovered in 1907 as bakelite and later in the 1960s it was known as polyethylene, which is cheap to make and their single use product plastic bag was a robust and easier option instead of paper or cloth bags. The accumulation of million tons of plastics in the ocean is known as microplastic, which will take 500 years to degrade completely.

Nearly all plastic is derived from materials (like ethylene and propylene) that are made from fossil fuels (mostly oil and gas). The whole process of extracting and transporting those fuels in order to manufacture plastic creates billions of tons of greenhouse gases.

Plastic has already been a nuisance that is difficult to degrade, releases carbon dioxide and other potent greenhouse gases. The unmanaged plastics that winds up in our ocean is a climate threat apart from the emissions caused by the plastic landfilling, incineration and recycling in addition to the human, environmental and economic problems it causes.

The ocean plays a major role in carbon sequestration in which the microscopic plants and creatures considered phytoplankton and zooplankton play out a basic assistance for human endurance. The critters present in ocean surface have a hunger for scrumptious plastic pieces and the petrochemicals present in them but excessive junk plastic meddles with their digestion, conceiving capacities and endurance; therefore, it affects their potential ability as a carbon sink. This will cause an impact on climate change. The plastic that covers the land mass will also cause damage to the soil and the nutrient uptake capacities of plants are hampered.

The appearance of the COVID-19 pandemic has upgraded the intricacies of plastic waste administration. Disturbing instances of contamination have exceeded the use of personal protective equipment (PPE) (containing a substantial extent of plastic) as the most solid and reasonable defence against disease and transmission of the infection (Herron et al., 2020). The expanded interest for single-use PPE by medical practitioners and other medical warriors whilst mandating the use of masks to safeguard oneself from the harmful contamination has changed the elements to rapid plastic usage. The prevalence of single-use plastics over other alternatives has shifted the consumer demand in support of plastic packaging and single-use plastic bags (Scaraboto et al., 2019). Also, public lockdowns and home isolate orders by the government have emancipated an expanded dependence on online purchase of food and other fundamental goods which has caused a plausible expansion in plastic packaging waste generation (Scaraboto et al., 2019; Singh, 2020; Staub, 2020).

The enormous demand in the use of PPE kits consisting of gloves, face masks and gowns, by healthcare workers, has increased the disposal after single use; therefore, it generates an enormous amount of plastic waste (WHO, 2016). Apart from that the protective equipment used by the healthcare staff and infected patients inside an ambulance including hoods, masks, gloves and gowns, during their transfer to healthcare facilities, are generally disposed of post-transport (Higginson et al., 2020). The humongous generation of biomedical waste from laboratory studies and testing has contributed to considerable increase in the proportion of plastics.

As per the World Wide Fund (WWF) report, even 1% improper disposal of masks would accumulate to 10 million masks per month, which will be polluting the environment (Italy WWF, 2020). The face masks imported from China are made of multiple layers of different polymers, thereby making it difficult to recycle (Monella, 2020).

As per the study by Klemeš et al. (2020), there was a steep rise in the infectious medical waste in China during COVID-19 outbreak and an increase from 40 tons waste per day to 240 tons per day was observed.

NATURAL RESOURCES CONTROLLING THE TEMPERATURE

As per the study conducted by the Intergovernmental Panel on Climate Change (IPCC), the oceans uptake around 90% of the excessive heat generated due to human intervention leading to global warming. The coral reef of Australia has witnessed the adversities of climate change and global warming. Half of earth's oxygen is made by coral reefs and it absorbs nearly one-third of the carbon dioxide which is generated from burning fossil fuels. As there will be an increase in temperature, mass coral bleaching will occur along with outbreaks of infectious diseases that became evident with the wide spread of novel coronavirus around the globe. The change in temperature will raise the sea level apart from changes in the frequent intensity of tropical storms, and thus alter ocean circulation patterns. The bleaching or discolouration in the corals occurs due to increase in temperature, which forces the corals to expel the symbiotic algae living in their tissues, responsible for their colour eventually leading to death. The death of coral will lead to loss of marine biodiversity of the ocean. According to a study reported in Global Oceanic Environmental Survey in 2021, the widespread use of sun cream/sun blockers with a composition of oxybenzone which actually works by changing the ultraviolet (UV) rays to a less energetic wavelength will be safer for human skin. The change in wavelength releases free radicals that are harmful for the corals, algae and planktons. The sun blockers form a thin layer of film on the surface of the water, giving it a shining emulsion-like appearance. The presence of micro plastics in the sea gets attached to these tiny emulsions to form a toxic compound which is engulfed by the planktons and slowly it makes its presence in our food chain.

According to the latest survey conducted by Global Oceanic Environmental Survey, the oceans play a very detrimental role in biological carbon sequestration as they sequester four times more carbon than terrestrial and half of all the human activity-generated carbon. It is more than the amount reported by IPCC (Pearce et al., 2018).

Table 3.2 gives a clear picture of the role played by oceans and the need to conserve and restore them for fostering their vital part in safeguarding the global climate change (www.ctbto.org/map/).

MICROPOLLUTANTS IN ENVIRONMENT

Plastics in its various forms like microplastics (Huang et al., 2020) and non-plastics have caused a wide impact on the natural biodiversity. The rising temperature along with an acidic environment makes it difficult for the coral inhabitants to cause break in disease spread, extreme storms or an influx of pollution created by microplastic. Corals constitute a silent but impactful entity in saving our planet. The amount of deterioration caused by human activities to the environment and its relevance to climatic change often makes us forget that these artificial entities roaming in space are getting accumulated in outer space as space debris. They are often negligent in the present scenario but eventually their giant form will create a no-go zone for the spaceships, then maybe it will be a cause of concern. These debris were not a concern till a few years ago but with the current alterations in the climate their omnipresence has been taken into consideration though climate change does occur throughout the atmosphere system. The troposphere solicits the trend of global warming, the other layers consisting of stratosphere, mesosphere and thermosphere have been cooling with a complicated trend patterns in the ionosphere (Cnossen, 2012; Laštovička et al., 2012; Qian et al., 2011), and this cooling invariably causes contraction in upper atmosphere and creates friction that eventually drags the space junk back to earth. There is a need for long-term changes in

TABLE 3.2
The Global Carbon Budget (GCB) 2020 Data Given by Global Oceanic Environmental Survey (GOES)

Carbon flux, gigatons of Carbon as C.	Global Carbon Budget 2020 data		GOES 2021 data		GOES 21 Carbon flux if we restore ocean productivity by eliminating pollution	
	Carbon to atmosphere	Carbon sequestered	Carbon to atmosphere	Carbon sequestered	Carbon to atmosphere	Carbon sequestered
Carbon from the burning of fossil fuels	9.4	-	9.4	-	7.7	-
Terrestrial ecology sequestration	-	3.4	-	0.6	-	1.2
Ocean ecosystems	-	2.5	-	2.5	-	5.0
Silicate mineral absorption	-	-	-	1.7	-	1.7
Volcanoes	0.1	-	0.1	-	0.1	-
Land use change (e.g. burning trees)	1.6	-	1.6	-	1.6	-
Total carbon flux	11.1	5.9	11.1	5.9	9.4	9.4
Total Carbon to atmosphere every year gigatons of C.	5.2					

upper atmosphere climate to predict our future state. These long-term predictions are necessary for various practical implementations, such as planning for satellite missions, management of the risks of space debris (e.g. Lewis et al., 2011), the review of long-term climate monitoring with satellite-based data (e.g. Scharroo & Smith, 2010).

The pollution in the form of gases and particles is evident in the exhaust clouds which are formed when the rockets go for the second stage. The rocket is made of light metal such as aluminium oxide and a single rocket motor can spread billions of particles of the metal which forms a cloud that lingers for around two weeks after the rocket has been fired, i.e. before dispersing and re-entering the atmosphere. The particles present a significant threat of surface erosion and contamination to spacecraft. Orbital debris further interferes with the scientific study and observation as combination of by-products from second stage firings – gases, small solid particles and space-glow will often affect the accuracy of scientific data. These space debris also contaminate the stratospheric cosmic dust collection experiments or even interfere with the debris tracking processes. Scientists now have difficulty in the prediction of the presence of debris as the space litter has aggravated the problem. With the increase in debris the amount of light reflected by them will vehemently make progression, therefore causing light pollution which will further interfere with astronomical observation. The debris will also disrupt the smooth reception of radio telescopes and therefore give distorted photographs that will affect the scientific results and predictions. These waste though look insignificant now will have a greater impact in the coming years.

Apart from the cosmic waste there are various forms of waste that were insignificant earlier but have taken prominent roles now. The contemporary waste doesn't mention their role but the predominance of antibiotic resistance occurring due to untreated disposal of pharmaceutical waste. According to the World Health Organization (WHO), antimicrobial resistance (AMR) and change in climate are some of the hotspot areas for global health threats.

The epidemiological study published in *Nature Climate Change* suggests that with an increase in local average minimum temperature by 10°C there is an association found with 4.2 percent, 2.2 percent and 3.6 percent increase in antibiotic resistant strains of *Escherichia coli*, *Klebsiella pneumoniae* and *Staphylococcus aureus*, respectively. Epidemiologists from the USA and Canada found higher local temperatures correlated with a higher degree of antibiotic resistance in common bacterial strains. With the present precarious situation the uses of antibiotics have increased and their proper disposal and treatment have already taken backstage as the pandemic has restricted the processing of these wastes. The accumulation of antibiotics in the environment will lead to the formation of antimicrobial resistance (AMR). As per the study of Center for Disease Dynamics, Economics & Policy (2015), globally it has already caused so many deaths and by 2050 it is projected to cause 200,000 deaths per year in India. The overuse in agriculture for livestock, and also rampant use by humans, has aggravated the problems. Industrial agriculture relies heavily on the widespread use of antimicrobials for livestock farms for therapeutics (Woolhouse et al., 2015), prophylactics and controversially growth promotion. Rodríguez-Verdugo et al. (2020) suggested that small doses of antibiotics from urine, faeces, manure and pharmaceutical waste are being released into the environment through rivers, lakes and soil. According to Andersson and Hughes (2014), these sub-lethal doses allow for antimicrobial resistance to occur as they do not reach "cidal" concentrations.

The increase in temperature has played a major role in increasing the antibiotic resistance for three common pathogens prevalent in the world, i.e. *Escherichia coli, Klebsiella pneumoniae* and *Staphylococcus aureus*.

The study reported in Earth.org (https://earth.org/climate-change-linked-to-antibiotic-resistance) depicted that the southern European countries with minimum temperatures that were 10°C warmer due to climate change –such as Spain, Portugal, Romania and Italy – recorded a more rapid rise in antibiotic resistance over time than cooler northern European countries such as Sweden, Finland and Norway. The increases in resistance ranged from 0.33% to 1.2% per year, even after accounting for factors like local population density and local patterns of antibiotic use. They further gave the information that *in vitro* experiments provide evidence where bacterial growth increases at warm temperatures that facilitate transmission of resistant strains. Higher temperatures can help in enhancing the transfer of antibiotic resistance genes between bacteria.

Lockhart et al. (2017) and Casadevall et al. (2019) hypothesized that *Candida auris*, which previously existed as a plant saprophyte, has arisen from climate change and therefore gained thermo-tolerance and salinity tolerance from the effects of climate change on the wetland ecosystem. It was initially isolated from a human ear in 2009 and hence has been associated with human disease in many countries and thus has exhibited non-susceptibility to antifungal agents (Casadevall and Pirofski, 2019). The reformed thermo-tolerant *C. auris* was proposed to come into existence through interaction between birds and humans. Migration of humans led to the emergence of *C. auris* into urban healthcare environments in which AMR and infection control issues have arisen.

C. auris is the only *Candida* species that has isolates which have shown to be resistant to all four classes of antifungal drugs, and therefore created higher risks for clinical infections and breakthrough infections during antifungal treatment and prophylaxis (Pristov and Ghannoum, 2019). These infections have the potential to result in significant morbidity and mortality which could be on the rise as climate change continues.

IMPACT OF TEMPERATURE AND GLOBAL REPERCUSSIONS

The increase in temperature might occur due to various factors like anthropogenic activities where nuclear testing has emancipated the release of harmful chemicals and radioactive materials in the environment. Atmospheric nuclear weapons testing therefore involves the release of considerable amounts of radioactive materials directly into the environment and causes the largest collective dose from man-made sources of radiation (UNSCEAR, 2000). Prăvălie (2014) postulated that Nevada desert in the USA is the major region where 44% of all the nuclear tests world-wide were conducted. The environmental consequences are related to atmospheric contamination with radioactive isotopes (especially ^{14}C and ^{137}Cs) following the atmospheric nuclear tests conducted during 1951–1963. Simon et al. (2006) gave the report of negative impacts of atmospheric contamination from radionuclides, ^{131}I and ^{133}I, which were later transferred to the biosphere mainly through rainfalls. The radionuclide ^{131}I is reported to be the main causative agent behind the increase in thyroid cancer in the USA as it was released in huge amounts during the nuclear testing conducted from 1951 to 1958 (Hundahl, 1998; Gilbert et al., 1998, 2010).

The isotope ^{14}C that is released into the atmosphere during nuclear tests, gets integrated into the CO_2, and therefore reaches the marine environment through the ocean–atmosphere gas exchange, or the biosphere through the process of photosynthesis.

In terms of human exposure (Figure 3.1), the increase in the incidence of thyroid cancer in many areas of the world (strongly affected by the radioactive contamination with the ^{131}I radionuclide) is the worst consequences of nuclear testing

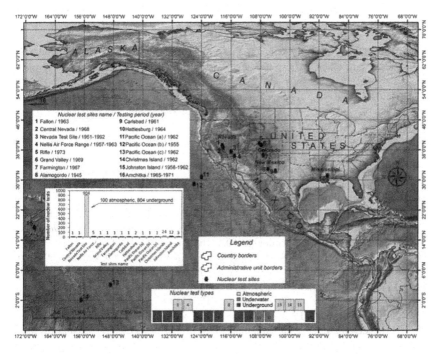

FIGURE 3.1 The spatial distribution, type, the number and the period/year of nuclear tests conducted by the United States during 1945–1992 (data processing from www.ctbto.org/map/) (Prăvălie, 2014).

and the main reason behind the epidemiological surge of diseases and climate change (Dryden, 2021).

REFERENCES

1. Hussain, M. and Mumtaz, S., 2014. E-waste: impacts, issues and management strategies. *Rev Environ Health*, *29*(1–2), 53–58.
2. Baldé, C.P., Forti, V., Gray, V., Kuehr, R. and Stegmann, P., 2017. *The global e-waste monitor 2017: quantities, flows and resources*. United Nations University, International Telecommunication Union, and International Solid Waste Association.
3. Zolnikov, T.R., Furio, F., Cruvinel, V. and Richards, J., 2021. A systematic review on informal waste picking: occupational hazards and health outcomes. *Waste Manag*, *126*, 291–308.
4. Li, W. and Achal, V., 2020. Environmental and health impacts due to e-waste disposal in China – a review. *Sci Total Environ*, *737*, 139745.
5. Lu, C., Zhang, L., Zhong, Y., Ren, W., Tobias, M., Mu, Z., et al., 2015. An overview of e-waste management in China. *J Mater Cycles Waste Manag* , *17*(1), 1–12.
6. Joon, V., Shahrawat, R. and Kapahi, M., 2017. The emerging environmental and public health problem of electronic waste in India. *J Health Pollut*, *7*(15), 1–7.
7. Kumar, S., Garg, D., Sharma, P., Kumar, S. and Tauseef, S.M., 2018. Critical analysis and review of occupational, environmental and Health issues related to inadequate disposal of E-waste. In *Intelligent communication, control and devices* (pp. 473–484). Singapore: Springer.
8. Amoyaw-Osei, Y., Agyekum, O., Pwamang, J., Mueller, E., Fasko, R. and Schluep, M., 2011. Ghana e-waste country assessment: SBC e-waste Africa project. Coordinated by the Basel Convention, March.
9. Rautela, R., Arya, S., Vishwakarma, S., Lee, J., Kim, K.H. and Kumar, S., 2021. E-waste management and its effects on the environment and human health. *Sci Total Environ*, *773*, 145–623.
10. United Nations Environment Programme (UNEP) Press Release., 2019. UN report: time to seize opportunity, tackle challenge of e-waste.
11. Manish, A. and Chakraborty, P., 2019. *E-waste management in India: challenges and opportunities*. TERI.
12. https://earthobservatory.nasa.gov/features/SeaIce.
13. Herron, J.B.T., HayDavid, A.G.C., Gilliam, A.D. and Brennan, P.A., 2020. Personal protective equipment and Covid-19 – a risk to healthcare staff? *Br J Oral Maxillofac Surg*, *58*, 500–502. https://doi.org/10.1016/j.bjoms.2020.04.015.
14. Scaraboto, D., Joubert, A.M. and Gonzalez-Arcos, C., 2019. Using lots of plastic packaging during the coronavirus crisis? You're not alone. *The Conversation*, *668*, 1077–1093. https://theconversation.com/using-lots-of-plastic-packaging-during-the-coronavirus-crisis-you're-not-alone-135553.
15. Singh, S.H., 2020. Lockdown helps bring back plastic bags. *The Hindu*. www.thehindu.com/news/national/telangana/lockdown-helps-bring-back-plastic-bags/ar- ticle31381638.ece.
16. Staub, C., 2020. Country lockdowns bring 'unprecedented implications'. *Resource Recycling*. https://resource-recycling.com/plastics/2020/04/01/country-lockdowns-bring-unprecedented-implications/.

17. Higginson, R., Jones, B., Kerr, T. and Ridley, A.-M., 2020. Paramedic use of PPE and testing during the COVID-19 pandemic. *Journal of Paramedic Practice, 12,* 221–225.
18. Italy WWF., 2020. In the disposal of masks and gloves, responsibility is required. *WWF International.* www.wwf.it/chi_siamo/organizzazione/.
19. Klemeš, J.J., Fan, Y. Van, Tan, R.R. and Jiang, P., 2020. Minimising the present and future plastic waste, energy and environmental footprints related to COVID-19. *Renew Sust Energ Rev, 127,* 109883. https://doi.org/10.1016/j.rser.2020.109883.
20. Monella, L.M., 2020. Will plastic pollution get worse after the COVID-19 pandemic? *Euronews.* www.euronews.com/2020/05/12/will-plastic-pollution-get-worse-after-the-covid-19-pandemic.
21. WHO., 2016. *Personal protective equipment for use in a filovirus disease outbreak: rapid advice guideline,* Geneva: WHO.
22. Dryden, H., 2021. Regenerate nature, our best hope to reverse climate change. *Our best hope to reverse climate change* (February 26, 2021).
23. Huang, W., Chen, M., Song, B., Deng, J., Shen, M., Chen, Q., et al., 2020. Microplastics in the coral reefs and their potential impacts on corals: a mini-review. *Sci Total Environ,* 143112.
24. Cnossen, I., 2012. Greenhouse gases emission, measurement and management. In G. Liu (ed.), *In tech* (pp. 315– 336). Croatia: Rijeka.
25. Laštovička, J., Solomon, S.C. and Qian, L., 2012. Trends in the neutral and ionized upper atmosphere. *Space Sci Rev, 168*(1–4), 113–145.
26. Qian, L., Laštovička, J., Roble, R.G. and Solomon, S.C., 2011. Progress in observations and simulations of global change in the upper atmosphere. *J Geophys Res, 116,* A00H03.
27. Lewis, H.G., Saunders, A., Swinerd, G. and Newland, R.J., 2011. Effect of thermospheric contraction on remediation of the near-Earth space debris environment. *J Geophys Res, 116,* A00H08.
28. Scharroo, R. and Smith, W.H.F., 2010. A global positioning system-based climatology for the total electron content in the ionosphere. *J Geophys Res, 115,* A10318.
29. https://ota.fas.org/reports/9033.pdf.
30. Center for Disease Dynamics, Economics & Policy., 2015. *State of the world's antibiotics.* Washington, DC: Center for Disease Dynamics, Economics & Policy.
31. https://earth.org/climate-change-linked-to-antibiotic-resistance/.
32. Pearce, W., Mahony, M. and Raman, S., 2018. Science advice for global challenges: learning from trade-offs in the IPCC. *Environ Sci Policy, 80,* 125–131.
33. Woolhouse, M., Ward, M., van Bunnik, B. and Farrar, J., 2015. Antimicrobial resistance in humans, livestock and the wider environment. *Philos Trans R Soc Lond Ser B Biol Sci, 370,* 20140083.
34. Lockhart, S.R., Etienne, K.A., Vallabhaneni, S., Farooqi, J., Chowdhary, A., Govender, N.P., et al., 2017. Simultaneous emergence of multidrug-resistant *Candida auris* on 3 continents confirmed by whole-genome sequencing and epidemiological analyses. *Clin Infect Dis, 64,* 134–140.
35. Casadevall, A., Kontoyiannis, D.P. and Robert, V., 2019. On the emergence of *Candida auris*: Climate change, azoles, swamps, and birds. *mBio, 10,* e01397–19.
36. Pristov, K.E. and Ghannoum, M.A., 2019. Resistance of Candida to azoles and echinocandins worldwide. *Clin Microbiol Infect, 25,* 792–798.

37. Casadevall, A. and Pirofski, L.A., 2019. Benefits and costs of animal virulence for microbes. *mBio*, *10*, e00863–19.

38. Rodríguez-Verdugo, A., Lozano-Huntelman, N., Cruz-Loya, M., Savage, V. and Yeh, P., 2020. Compounding effects of climate warming and antibiotic resistance. *iScience*, *23*, 101024.

39. Andersson, D.I. and Hughes, D., 2014. Microbiological effects of sublethal levels of antibiotics. *Nat Rev Microbiol*, *12*, 465–478.

40. UNSCEAR., 2000. Report to the general assembly (Annex C—Exposures to the public from man-made sources of radiation).

41. Prăvălie, R., 2014. Nuclear weapons tests and environmental consequences: a global perspective. *Ambio*, *43*(6), 729–744.

42. Hundahl, S.A., 1998. Perspective: National Cancer Institute summary report about estimated exposures and thyroid doses received from iodine 131 in fallout after Nevada atmospheric nuclear bomb tests. *CA Cancer J Clin*, *48*(5), 285–298.

43. Gilbert, E.S., Tarone, R., Bouville, A. and Ron, E., 1998. Thyroid cancer rates and 131I doses from Nevada atmospheric nuclear bomb tests. *J Natl Cancer Inst*, *90*(21), 1654–1660.

44. Gilbert, E.S., Huang, L., Bouville, A., Berg, C.D. and Ron, E., 2010. Thyroid cancer rates and 131I doses from Nevada atmospheric nuclear bomb tests: an update. *Radiat Res*, *173*, 659–664. doi: 10.1667/RR2057.1.

45. Ackah, M., 2019. Soil elemental concentrations, geo accumulation index, non-carcinogenic and carcinogenic risks in functional areas of an informal e-waste recycling area in Accra, Ghana. *Chemosphere*, *235*, 908–917.

46. Liu, J., Xu, X., Wu, K., Piao, Z., Huang, J., Guo, Y., et al., 2011. Association between lead exposure from electronic waste recycling and child temperament alterations. *Neurotoxicology*, *32*(4), 458–464.

47. Simon, S.L., Bouville, A. and Land, C.E., 2006. Fallout from nuclear weapons tests and cancer risks. www.cancer.gov/PublishedContent/Files/cancertopics/causes/i131/fallout.pdf. Retrieved January 25, 2013.

48. Ha, N.N., Agusa, T., Ramu, K., Tu, N.P., Murata, S., Bulbule, K.A. et al., 2009. Contamination by trace elements at e-waste recycling sites in Bangalore, India. *Chemosphere*, *76*(1), 9–15.

4 Climatic Change in Context with Health

HUMAN HEALTH DIRECTLY PROPORTIONAL TO CHANGE IN CLIMATE

The betterment of human health is related to climate and with time it is becoming evident that the spread of diseases is severe when there is a climatic change. The depletion of natural resources has aggravated the problem in the present context.

Through this chapter you will get a glimpse of the impact of climate change on human health.

The foundation of the climate system depends on several interconnected subsystems: the atmosphere, the hydrosphere (rivers, lakes and oceans), the cryosphere (ice and snow), the lithosphere (soils) and the biosphere (ecosystems).

According to the Intergovernmental Panel on Climate Change (IPCC, 2007), there are numerous factors that have contributed to the climatic conditions, such as the volcanic eruptions, atmospheric composition, orbiting of Earth around the Sun and also the solar activity. The anthropogenic activities leading to global warming are ten times more than natural factors in the last few decades and are continuously contributing with the advancing years. Other factors that affect the climate are urbanisation, deforestation, forest fires, rate of emissions of greenhouse gases (GHGs) and industrialisation. Some of the important consequences of global warming lie in climate change, impacts on sea level rise, natural ecosystem, biodiversity, agriculture, forestry and food productivity, aquatic ecosystem, flora and fauna, glaciers and the detrimental factor of health.

The average weather condition of an area is termed as the climate of a region. The region includes rainfall, temperature variation, humidity and wind. The climatic condition of an area is affected by topography, longitude, latitude, Sun–Earth's axis, proximity to sea and oceans, wind directions, and temperature differences between land and sea. The change in climate is often termed as global warming and it refers to the gradual increase in average temperature on Earth's surface.

According to scientific consensus there is a continuous increase in the global temperature from 0.4 to 0.8°C during the past century and the cause for this escalation in temperature is attributed to the emission of carbon dioxide (CO_2) and other

DOI: 10.1201/9781003120629-4

greenhouse gases (GHGs) in the atmosphere. The combustion of fossil fuels gives rise to the increased volume of CO_2 and other GHGs.

With the increase in population there is an immediate surge for land and in order to cater the anthropogenic demands the forest was cleared. Apart from this agricultural activity has also significantly contributed to global warming. The most vital factor responsible for climate change lies in the increase in the concentration of the GHGs and CO_2 in the atmosphere. The economic upliftment of the nation depends on the industrial activities such as energy, industry, transport, land use and they rely heavily on the use of fossil fuel. 77 percent of global warming is attributed to CO_2 apart from methane generated by agriculture and rising land clearance, leading to deforestation. Stern (2006) defined the increase in CO_2 concentration level to nearly 100 parts per million (ppm). The present data elucidate that 2–3 ppm of CO_2 is the global emissions. The increased global warming is predicted to show its impact on working people and productivity by 2045.

Global warming and climate change have a very strong interrelationship in environment. The capacities of the GHGs to entrap the solar heat within the atmosphere have a detrimental impact on natural habitats, health and also agriculture. According to studies there was an abrupt end to the ice age 7,000 years ago. After the end of the ice age there was the beginning of the modern climatic era of human civilisation. There is an increase in sea level as the glaciers and polar ice caps are melting due to global warming, which will lead to severe floods in some areas, droughts and other natural calamities. The Fourth Assessment Report of IPCC (2007) gave a clear flowchart for the global warming scenario in the 21st century. The average global temperature as predicted by scientists could increase the temperature between 1.4°C and 5.8°C in the coming years. Despite best efforts global warming has influenced the whole world and the variation in climate is now more evident.

Precise researches conducted on climate change events during the past 150 years have mentioned about the increase in the atmospheric temperature over the world and the warming of the earth has occurred in two phases: the first phase started from 1919 to 1940, with an average gain of temperature of 0.35°C, and the second phase started from 1970 to present, with a 0.55°C gain in temperature. The oceans have shown increased heat with the top 700 meters of ocean water showing warming random increase.

In the last 25–30 years, the records of climate change showed the warmest time in the past five centuries. The increase in global warming has caused the rise in average temperatures of the oceans, rising of the sea levels, melting of ice glaciers and thinning of the snow cover in the Northern and Southern Hemispheres. The prehistoric evidence showed the climate change that is evident in tree rings, ocean sediments, coral reefs and layers of sedimentary rocks. This evidence showed that the current rate of global warming is faster than the average rate of ice-age recovery warming. The Earth's climate responds to atmospheric changes by the levels of GHGs. The Earth's average surface temperature has shown an increase of nearly 2.0°F (1.1°C) since the last century. The maximum amount of significant changes in global warming was observed in 2001. The mass of ice sheets observed in Greenland and Antarctic has declined. The scientific study observed by National Aeronautics and Space Administration's (NASA) Gravity Recovery and Climate Experiment showed that Greenland has nearly lost 150–250 cubic kilometres (36–60 cubic miles) of ice per year between 2002 and 2006, while Antarctica has lost nearly 152 cubic kilometres (36 cubic miles) of ice between 2002 and 2005. The length and thickness of Arctic Ice has also decreased over the years.

Burning of fossil fuels releases huge amounts of GHGs into the atmosphere, mainly CO_2, which is the main source of global warming. Other human activities include the stubble burning for crop remains in order to clear the land and the rapid deforestation which have also significantly contributed to the release of GHGs. The atmospheric concentration of CO_2 has increased from pre-industrial era to present. Global levels of CO_2 crossed the symbolic 400 parts per million (ppm) benchmark in 2015. It is significant to note that the atmospheric concentration of CO_2 rose to above 300 parts per million (ppm) between the advent of human civilisation approximately 10,000 years ago and 1900. But at present times, its concentration in the atmosphere has reached to about 400 ppm, which is a level that has never been reached in more than 400,000 years in the Earth's history and therefore is the main reason for global warming. An increasing rate of global warming has taken place over the last 25–30 years.

The consequences of global warming will have a significant impact on the natural environment, food supply, water resources, land, infrastructure and the increase in weather events, and socio-economic activities. Climate change will have a deep impact on natural biodiversity, agricultural productivity, food security, human health, coastal zone and aquatic ecosystem.

The increase in temperature will cause shifts in crop growing seasons and cropping patterns that will affect the food security. It will also cause melting of glacial ice, leading to soil erosion and downstream flooding. Temperature fluctuations will affect the rate of extinction for natural habitats and indigenous species.

A great number of people depend on agriculture for their livelihood and they have to bear the brunt of climate change. With the increasing population there is a demand for adequate food supply and it has been estimated that by 2050 there will be an upsurge of 2.2 billion people. With the changing climate it will have a greater impact on the environment. The frequency of intense rainfall has increased in the present scenario, making low lying areas more prone to floods, landslides that are more frequent in hilly regions and debris flow in many regions. In addition, the total amount of precipitation has declined in the West, Southwest and Southeast regions of the United States.

With the change in season there has been frequent rainfall witnessed in the United States in the last six decades. From the 1950s, the record high temperature event in the United States has increased and low temperature events are decreasing. The level of acidity in ocean water has increased by nearly 30 percent since the inception of the Industrial Revolution and the anthropogenic emission of CO_2 in atmosphere has added fuel to the situation.

There has been an increase of about 2 billion tons of CO_2 absorbed in the upper layer of the ocean. The rising temperature in the sea levels in coastal areas will create greater storm surge, inundation and damage to coastline due to high waves that are more prevalent in countries with small islands and low lying deltas. According to the IPCC reports, the probable impacts of climate change are projected due to global warming on some of the major sectors, such as coastal areas, sea level rise, biodiversity, glaciers, floods, agriculture and human health.

The increasing sea level due to global warming will escalate the loss of land and people as there will be permanent inundation in the coastal areas. According to IPCC Fourth Assessment Report, South and Southeast Asian countries will be severely affected due to adverse climatic conditions as most of the countries in that region depend solely on natural resources and agriculture. Countries like Bangladesh, India, Vietnam and Sri Lanka that have larger coastlines will also face the consequences of climate change. The frequent weather events like rising sea level and increasing cyclonic activity have become more frequent in the Bay of Bengal and Arabian Sea, leading to an increase in waterborne diseases (Kopprio et al., 2020). The climatic impacts will threaten the livelihood of the marginal people living in rural and coastal areas. The change in climatic conditions will create adverse situations leading to severe weather events, leading to wildfires that threaten natural biodiversity, habitats, homes and lives, and heatwaves may contribute to human deaths.

The IPCC reported that various studies have projected an increase in sea level rise that would flood the coastal residential areas of millions of people who are living in the low lying areas of South, Southeast and East Asia, such as in Sri Lanka, Vietnam, Bangladesh, India and China (Wassmann et al., 2004; Stern, 2006; Cruz et al., 2007). Ongoing sea level rise has displaced thousands of people in the Sunderbans.

The CO_2 concentration accounts for a major proportion of GHGs as they are increasing at a rapid rate which has led to higher growth and plant productivity due to higher photosynthesis rate but offsets this effect as it leads to increased crop respiration rate and evapotranspiration, increased pest infestation, weed flora and reduced crop duration. Change in climate affects the microbial population and enzymatic activities in soil (Malhi et al., 2021). Climatic disaster will hamper the economy of the countries that majorly depend on agriculture, such as Sri Lanka, Vietnam, Bangladesh, India and China. India's gross domestic product (GDP) will decline to about 9%, with a 40% reduction in the production of the major crops. A temperature increase of 2°C in India is projected to displace 7 million people in the near future. The change in the intensity of rainfall events and the break cycles of the monsoon along with critical temperatures would significantly change the cropping pattern and yield of crops.

The effects of climate change on agriculture will vary according to locality, but it is projected that there will be 15–30 % decline in the crop productivity of most cereals and rice across the region (Lee et al., 2008).

The global climate will contribute to the extinction of over 1% of the world's aquatic life and will also accelerate the species extinction and thereby a loss of many vulnerable and endangered flora and fauna.

THE IMPACT OF CARBON DIOXIDE ON NATURAL ECOSYSTEM

The increasing levels of CO_2 have a direct impact not only on the natural ecosystem and rich biodiversity but also on temperature and rainfall patterns. Natural resources such as medicinal plants, timber products and fisheries support the commercial aspect of the region. The degradation and loss of natural ecosystems pose a serious threat to the economic, social and cultural areas of the region. The survival of many species has already been threatened due to land-use changes, land degradation, loss of biodiversity, over-exploitation of water resources and other natural resources, and contamination of inland and coastal water bodies.

The increasing global warming will have an effect on sea level rise, increasing sea surface temperature as well as acidification of the oceans and other water bodies. It will also have a greater impact on the natural habitat like coral reefs, mangrove forest and reduction in fish. According to the IPCC Fourth Assessment Report, 1997–1998 El Nino event had a widespread effect in the bleaching of coral reefs in the Southeast Asian region including Indonesia, Cambodia, Malaysia and Thailand.

The Himalayan ecosystem is affected by sudden glacial lake outburst, floods and flash floods. The increasing sea level rise and sea surface temperature are the major climate change-related stresses on coastal ecosystems.

According to the World Bank's 2014 report released by the IPCC (2014), people living in coastal regions of Asia are likely to lose their shelter, homes and settlements due to flooding and famine. The major coastal cities, ports and tourist resorts, commercial and small-scale fisheries, coastal agriculture and infrastructure development will impact the socio-economic aspect. Apart from these the increased salination of surface water and groundwater, wetlands and public health will have greater risks (IPCC Fourth Assessment Report).

The effect of global warming has made changes in the distribution of water around the world and the availability of water between the regions has been difficult. Many areas are dependent on the limited availability of groundwater and rainfall collection. Climate change has aggravated the situations and the challenges to water resource management were further exacerbated by the increase in sea level rise that leads to the intrusion of sea water into the available freshwater in the coastal areas.

Scientific assessments have shown the increase in water management costs and the increasing water stress for the rural marginal people due to changing patterns of run-off flows. Climate change indicators and impacts have deteriorated in 2020, which was recorded as the warmest year compared to three previous consecutive years. Besides temperature, droughts, heat, floods, cyclones and wildfires impact countries around the world.

The increase in sediments and reduction of in-flow from snow-clad rivers will create shortages for the water to be used in hydropower generation, urban supply and agriculture. The increasing demands for water to cater the needs of agricultural, industrial and hydropower sectors will put additional stress on available water resources. There is a limited availability of freshwater due to restricted seasonal availability of 75% annual rainfall which occurs during monsoons.

The supply of water is restricted by higher temperatures, changes in the river flow and occasional coastal flooding. The supply of water is minimal during the dry season. The melting of glaciers due to global warming plays an important role in the provision of water to the region and snowfields which are currently supplied with up to 85% of the dry season flow of the great rivers. Hydrological changes will be significant with the increasing run-off in river basins in response to increasing rainfall.

WITH INCREASE IN TEMPERATURE THE RESOURCES ARE AFFECTED

According to recent data from the IPCC Fourth Assessment Report, regional studies have shown a loss of $1.7 billion in the total cost of the water resources sector that projects a 2% increase in temperature and 2–4% increase in temperature due to water stress.

In the present context, the increasing temperature due to anthropogenic activity plays an important role in global climate change and also has an impact on hydrological cycle that helps in the continuous movement of water on earth. The water-holding capacity of the atmosphere is a function of the temperature and it increases by about 8% per degree Celsius (Trenberth, 1999).

The important GHG in the atmosphere is water vapour that not only allows the visible light to pass through but also absorbs the part of infrared radiation from the earth, and therefore retains the heat in the system. The study by Semenza et al. (2012) gave the introspection that water cycle increases the temperature and the evaporation, and therefore the atmospheric moisture content of both frequency and extreme precipitation.

Water has the capacity to absorb approximately 20 times more heat than the atmosphere (Levitus et al., 2012). More than 90% of Earth's warming is attributed to the ocean around the world. With the top layers of the ocean (the uppermost 700 metres) that contribute majorly to the increasing warming trend there is significant warming also shown by the deeper ocean (between 700 and 2,000 metres depth). The quantity and quality of freshwater is also affected by global climate change.

An increase in hot spells and heatwaves leads to a significant rise in droughts and excessive rainfall leads to flood. These unpredictable hydrological changes can affect waterborne diseases because their environmental exposure pathways are linked to local climate and weather conditions. Semenza, Herbst, et al. (2012) deciphered that some waterborne pathogens outside the human and animal host cannot multiply (e.g. Norovirus, *Cryptosporidium* and *Camplylobacter*), whereas some can survive in environment (e.g. *Vibrio, Salmonella*).

CLIMATE CHANGE AND DISEASE BURDEN

The current burden of diseases is due to climate-sensitive health outcomes and is not limited to diarrhoea, vector-borne diseases, malnutrition, deaths due to floods and landslides and cardiovascular diseases in cold waves and heatwaves.

In developing nations, the pathogens involved and the exposure pathways differ. The significance of cholera outbreak has been associated with heavy rains (Griffith et al., 2006); in Korea, hepatitis A outbreak is associated with the contamination of water source (Lee, C.S. et al., 2008); the enterovirus outbreaks in Taiwan has been linked to extreme precipitation (Jean et al., 2006). Due to the failure of the water treatment facility the presence of faecal coliform in water becomes evident after a heavy rainfall (Richardson et al., 2009). The occurrence of microbial nutrients and organic matter in water distribution system due to heavy rain run-off fosters the regrowth in treated water (LeChevallier et al., 1991).

The impact of climate change on the hydrological cycle and waterborne diseases has impacted the surveillance practices (Lindgren et al., 2012). Climate change adaptation strategies should be designed to improve the public health preparedness and facilitate the response to emerging threats from waterborne diseases and thereby should be designed to contain the human and economic costs. The negative consequences of the anticipated extremes of the hydrological cycle will be mitigated through adaptive strategies. Waterborne and vector-borne pathogens have closely correlated with the changing dynamics of water. With the changing dynamics of water these pathogens have become more prevalent. Interventions and guidance of public health are essential for both mitigation and monitoring the effects of climate change. The impact of climate change in developing nations is more prominent by the increase in epidemics of malaria, dengue and other vector-borne diseases (Martens, 1999). The attributable diarrhoea as a result of climate change is more prevalent in Southeast Asian countries such as Bhutan, Bangladesh, Nepal, India, Maldives and Myanmar in the year 2000. The impact of climate change on human health will be stronger as the increasing malnutrition and heat stress are making the human body more prone to diseases. The increase in temperature will have detrimental effects on health and lives along with associated economic and environmental activities.

As per the assessment of WHO, climate change will affect the social and environmental aspects of health, such as clean air, fresh and safe drinking water, sufficient food and secure shelter. By 2030 and 2050 climate change is expected to cause around 250,000 additional death/year from malnourishment, heat stress and diarrhoea. The current pandemic situation has been the one that has already aggravated the situation and death is evidently increasing with each passing day. The proper management of handling the pandemic has been the most difficult aspect in the present situation as the harmful effects of COVID-19 virus were unknown to mankind. One virus with its mutant has caused such havoc to mankind and if we couldn't handle the growing global warming the melting glaciers will release many more such microorganisms.

REFERENCES

1. Intergovernmental Panel on Climate Change., 2001. *Climate change 2001: synthesis report. A contribution of working groups I, II, and III to the third assessment report of the Intergovernmental Panel on Climate Change*, edited by R.T. Watson and the Core Writing Team (eds.). Cambridge: Cambridge University Press.
2. Trenberth, K., 1999. Conceptual framework for changes of extremes of the hydrological cycle with climate change. *Climate Change, 42*, 327–339.
3. Semenza, J.C., Caplan, J.S., Buescher, G., Das, T., Brinks, M.V. and Gershunov, A., 2012. Climate change and microbiological water quality at California beaches. *Ecohealth, 9*(3), 293–297.
4. Semenza, J.C., Herbst, S., Rechenburg, A., Suk, J.E., Höser, C., Schreiber, C. et al., 2012. Climate change impact assessment of food- and water borne diseases. *Crit Rev Environ Sci Technol, 42*, 857–890.
5. Griffith, D.C., Kelly-Hope, L.A. and Miller, M.A., 2006. Review of reported cholera outbreaks worldwide, 1995–2005. *Am J Trop Med Hyg, 75*, 973–977.
6. Lee, C.S., Lee, J.H. and Kwon, K.S., 2008. Outbreak of hepatitis A in Korean military personnel. *Jpn J Infect Dis, 61*, 239–241.
7. Jean, J., Guo, H., Chen, S., Liu, C., Chang, W., Yang, Y., et al., 2006. The association between rainfall rate and occurrence of an enterovirus epidemic due to a contaminated well. *J Appl Microbiol, 101*, 1224–1231.
8. Richardson, H., Nichols, G., Lane, C., Lake, I. and Hunter, P., 2009. Microbiological surveillance of private water supplies in England: the impact of environmental and climate factors on water quality. *Water Res, 43*, 2159–2168.
9. LeChevallier, M., Schulz, W. and Lee, R., 1991. Bacterial nutrients in drinking water. *Appl Environ Microbiol, 57*, 857–862.
10. Kopprio, G.A., Neogi, S.B., Rashid, H., Alonso, C., Yamasaki, S., Koch, B.P., et al., 2020. Vibrio and bacterial communities across a pollution gradient in the Bay of Bengal: unraveling their biogeochemical drivers. *Front Microbiol, 11*, 594.
11. Malhi, G.S., Kaur, M. and Kaushik, P., 2021. Impact of climate change on agriculture and its mitigation strategies: a review. *Sustainability, 13*(3), 1318.
12. Levitus, S., Antonov, J.I., Boyer, T.P., Baranova, O.K., Garcia, H.E., Locarnini, R.A., et al., 2012. World ocean heat content and thermosteric sea level change (0–2000 m), 1955–2010. *Geophys Res Lett, 39*(10), 1955–2010.
13. Lee, S.H., Heo, I.H., Lee, K.M., Kim, S.Y., Lee, Y.S. and Kwon, W.T., 2008. Impacts of climate change on phonology and growth of crops: in the case of Naju. *J Kor Geograph Soc, 43*(1), 20–35.
14. Lindgren, E., Andersson, Y., Suk, J.E., Sudre, B. and Semenza, J.C., 2012. Monitoring EU emerging infectious disease risk due to climate change. *Science, 336*(6080), 418–419.
15. Martens, P., 1999. How will climate change affect human health? The question poses a huge challenge to scientists. Yet the consequences of global warming of public health remain largely unexplored. *American Scientist, 87*(6), 534–541.

16. IPCC., 2007. *Climate change 2007: the physical science basis: summary for policy-makers* (pp.104–116). Geneva: IPCC.
17. Stern, N., 2006. *Stern review: The economics of climate change.*
18. Wassmann, R., Hien, N.X., Hoanh, C.T. and Tuong, T.P., 2004. Sea level rise affecting the Vietnamese Mekong Delta: water elevation in the flood season and implications for rice production. *Climatic Change, 66*(1), 89–107.
19. Cruz, R.V., Harasawa, H., Lal, M., Wu, S., Anokhin, Y., Punsalmaa, B., et al., 2007, Asia. Climate change 2007: impacts, adaptation and vulnerability. In M.L. Parry, O.F. Canziani, J.P. Palutikof, P.J. van der Linden and C.E. Hanson (Eds.), *Contribution of working group ii to the fourth assessment report of the Intergovernmental Panel on Climate Change* (pp. 469–506). Cambridge: Cambridge University Press.
20. Intergovernmental Panel on Climate Change (IPCC), 2014. *AR5 synthesis report: Climate change 2014.* Geneva: IPCC.

5 Deliquescing of Permafrost

Climate change has an unprecedented effect on the permafrost soil that leads to its melting. Permafrost is located below the upper layer of the soil and extends from 3 feet to 4,900 feet thick. The composition of permafrost includes rock, soil, sediments and varying amounts of ice that binds the elements together. Some permafrost has been frozen for centuries. Permafrost is the storehouse of the carbon-based remains of plants and animals that was transformed into ice before they could decompose. In scientific study it was estimated that the world's permafrost holds 1,500 billion tons of carbon, double the amount of available carbon in the atmosphere. Upon thawing of permafrost carbon dioxide (CO_2) and methane were released into the atmosphere. With the increase in temperature permafrost is transforming into a source for the global heat emission rather than storing the carbon. According to the United Nations (UN), permafrost thaw was one of the significant emerging environmental issues in 2019 (UN Environment, 2019).

THE ICE AGE

Ice that has been solid for thousands of years has preserved the potential virus and bacteria which have the capability to come back to life if they get favourable conditions. In a recent study it was found that *Bdelloid rotifers* came back to life after 24,000 years. Despite being minute they can survive in extreme conditions such as arid, frozen, without food and oxygen.

According to the study by Legendre et al. (2014), a giant variant of virus named *Pithovirus sibericum,* which is ten times larger than human immunodeficiency virus (HIV), can revive back and that it is still infectious enough to cause various diseases. Around 18.8 million square kilometres of northern soils permafrost have the capacity to hold 1,700 billion tons of organic carbon in the form of dead plants and animals. This is estimated to be four times more than the total carbon emitted by human activity (Tarnocai et al., 2009).

DOI: 10.1201/9781003120629-5

GLOBAL WARMING LEADING TO DISEASE PREVALENCE

Global warming has paved the way for the ancient viruses which were earlier harmful to humans to resurface again and cause havoc by rapid spread. In 2016, there was news that a 12-year-boy in Russia was infected by anthrax. There was an outbreak of anthrax and around 20 people were affected. The anthrax spores were released when there was a warm climate experienced in the Arctic Circle. This warm weather thawed the Siberian tundra and the carcass of anthrax-infected reindeer has released the spores.

These reindeers were infected during the Siberian plague which killed 2,500 reindeer and a boy due to anthrax infestation. Apart from the deadly anthrax there is some more research where US researchers have successfully revived the Spanish flu of 1918 which has killed millions of people from a fragment of a cadaver's lung which was frozen in the Alaska permafrost.

The fragments of smallpox were also discovered in the Siberian permafrost which was supposedly eradicated from the human habitat. The nomads used to bury the dead bodies near the river banks and thawing of permafrost will lead to their exposure to the surface.

Apart from the release of thousands of years of microorganisms there are other important factors that will be affected by melting of permafrost.

There will be an increase in the amount of greenhouse gases (GHGs) released leading to global climate issues. The following are few of the factors that will have impact on the climate:

(1) Surface albedo declination (initiated by melting of the Arctic ice cover),
(2) Water vapour increase in the atmosphere (by higher temperatures),
(3) Changes in the concentrations of the GHG in the atmosphere (initiated by the absorption of CO_2 in biomass and oceans),
(4) Emission of carbon (CH_4 and CO_2 generation from thawing permafrost).

THE WHITE PRISTINE GLORY OF UNCERTAINTIES

The mighty Himalayas have a large concentration of glaciers and prominent snow-fields. During winters there is an increase in snowfall. Melting of the snow during summers is an important source of water for many rivers. Melted water from the Himalayan glacier contributes a sizable portion of water to the main rivers like the Ganges, Brahmaputra, Indus and other rivers of South Asia. Climate change has a deep-seated impact on the Himalayas as the excessive melting of snow in other seasons will lead to a distribution run-off. The availability of freshwater for human use will be hampered as the unseasonal water increase will lead to flash floods that will affect the irrigation, hydropower generation, drinking water and other uses of water. The recent glacier burst that occurred in Uttarakhand in India in the most of February 2021 was due to the change in climate.

As per the Special Report on Ocean and Cryptosphere (SROCC) by the Intergovernmental Panel on Climate Change (IPCC), it was projected that glacier retreat and permafrost thaw will decrease the stability of glacier lakes. The excessive use of concrete cement structures instead of the traditional wood and stone masonry has generated the heat island impact in the mountain region. Climate change has altered the frequency and magnitude of natural calamities. The thawing of permafrost has transformed the Arctic. Huge amounts of GHGs like methane, CO_2 and nitrous oxide were released with the thawing of permafrost.

The release of GHGs in the Tundra region occurs in two ways: firstly, when permafrost thaws the microorganisms break down the organic matter and release methane and CO_2 into the atmosphere. Secondly in addition, research has also found that toxic petroleum wastes that were frozen in the pits also migrate towards the nearby freshwater ecosystem. These sumps were excavated by the oil and gas industry in the 1970s and 1980s and presently thawing with the change in climate.

Peat which consists of partly decayed vegetation concentrated in a water-saturated environment without oxygen, helps in preserving the permafrost from the impacts of climate change. Found in the low Arctic peat encompasses the whole active layer.

NATURE WAY OF STOPPING THE THAWING OF PERMAFROST

Impacts of thawing permafrost can be seen in various ways when the ground becomes unstable and can slump causing landslides, floods and coastal erosions. It harms the ecosystem giving rise to *Themokarst* lakes, which are formed from depression of the collapsed permafrost, and therefore are filled with melt water.

Simultaneous thawing will drain the lakes and wetlands completely, thereby destroying the biologically important resources.

Residues from landslides, muddy water and lakes will affect the plant life at the base of the food chain and also the creatures dependent on these.

The thawed permafrost will also affect the breeding pattern of Caribous and other Arctic species. Potential microorganisms capable of creating harmful diseases are presently frozen in the permafrost and with the change in climate along with thawing will release them into the atmosphere. The unknown microorganisms will significantly make their impact in the form of pandemic and epidemic.

The building or residential construction is difficult to solicit in the unstable ground and the heat generated from the buildings and pipelines will warm the permafrost in the long run, leading to thawing.

The study of the University of Edinburgh gives insights that permafrost thawing can be reduced by incorporating the plant communities, which include trees, shrubs and mosses, that play a vital role in restricting the temperature of soil. The shades of the plant protect the soil from harsh sunlight and the roots absorb the water, making it drier and better insulators. Mosses provide a layer of the soil which helps in insulation, sequestering the air and restricting the penetration of heat from

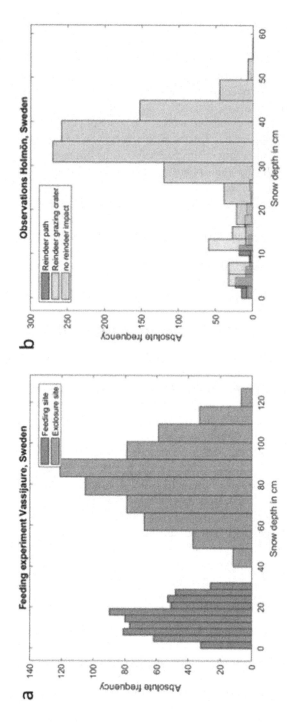

FIGURE 5.1　Histograms showing the snow depth observations at two locations in Sweden with/without reindeer impact (Zimov et al., 2012).

reaching the ground. After the death of the plant the remains help the permafrost by creating organic soils that act as insulators.

Herbivores also play an important role in preserving the permafrost by grazing and moving around during winter seasons. It was hypothesised that with their movement snow gets compacted and therefore decreases its thermal insulation efficiency. The mean annual soil temperature helps in freezing of the soil and hence prevents the thawing and also the degradation (Zimov et al., 2012).

Figure 5.1 shows the impact of reindeer (with/without) for the increase in snow depth in two areas of Sweden. The intervention of reindeer helps in refreezing of the top layers of permafrost after summer thaw and helps to keep it frozen for a longer time. The Arctic animals compact the snow when they graze, thereby packing it into a denser and thinner layer. This thinner snow layer attracts the cold Arctic air and helps in preserving the permafrost and curtails 4% of the warmth above the ground due to the insulating nature of the snow and also helps in thickening of the snow (Beer et al., 2020).

REFERENCES

1. UN Environment, 2019, *Frontiers 2018/2019 – emerging issues of environmental concerns*, Nairobi: United Nations Environment Programme.
2. Tarnocai, C. et al., 2009. Soil organic carbon pools in the northern circumpolar permafrost region. *Global Biogeochem Cycles, 23*, GB2023.
3. Legendre, M., Bartoli, J., Shmakova, L., Jeudy, S., Labadie, K., Adrait, A., et al., 2014. Thirty-thousand-year-old distant relative of giant icosahedral DNA viruses with a pandoravirus morphology. *Proc Nat Acad Sci, 111*(11), 4274–4279.
4. https://news.climate.columbia.edu/2018/01/11/thawing-permafrost-matters/.
5. https://timesofindia.indiatimes.com/india/climate-change-behind-uttarakhand-glac ier-burst-experts-feel/articleshow/80736594.cms.
6. https://e360.yale.edu/features/how-melting-permafrost-is-beginning-to-transform-the-arctic.
7. www.ed.ac.uk/sustainability/what-we-do/climate-change/case-studies/climate-research/can-plants-prevent-permafrost-thaw.
8. Zimov, S., Zimov, N., Tikhonov, A. and Chapin, F., 2012. Mammoth steppe: a high-productivity phenomenon. *Quat Sci Rev, 57*, 26–45.
9. Beer, C., Zimov, N., Olofsson, J., Porada, P. and Zimov, S., 2020. Protection of permafrost soils from thawing by increasing herbivore density. *Sci Rep, 10*(1), 1–10.

6 The Invincible Forging Ahead

CLIMATE CHANGE AND ITS IMPACT

The ever-changing climate has an impact on various aspects of the environment. The overuse of pharmaceuticals with the changing climate has invariably stressed the environment. In order to control the spread of coronavirus disease 2019 (COVID-19) there was widespread use of disinfectants and pharmaceuticals. These disinfectants were used alone or in combinations, i.e., alcohols, chlorine, formaldehyde, ortho-phthalaldehyde, hydrogen peroxide, acetic acid, phenolics and various quaternary ammonium compounds on the recommendation of public health authorities. Apart from the disinfectants there is mass production of pharmaceuticals, antibiotics in particular to constrain the bacterial infections (Watkinson et al., 2007). Over 250 million antibiotic prescriptions are written annually in the United States (US) alone. These antibiotics are not metabolised completely or eliminated from the body and 30–90% is excreted unchanged into the waste system (Watkinson et al., 2007).

ANTIBIOTIC RESISTANCE

The wastewater treatment process used traditionally can only remove 20–80% of pharmaceuticals and their metabolites; therefore, through direct or indirect disposal these antibiotics will be emitted in environments (Kovalova et al., 2012). The impact on the environment has become a worldwide concern as disinfectants and antibiotics are frequently detected in surface waters, ground waters, soils, sediments and wetlands with an amount up to 1 mg/L.

Many antibiotics were widely used to resist the COVID-19-induced inflammation and other diseases. The increasing misuse and use therefore has elevated the concentration and dose of disinfectants and antibiotics in the environment.

The non-target environmental organisms/microorganisms are also affected due to antibiotics or therapeutic drugs and disinfectants. Studies have shown the occurrence of disinfectants by-products (DBPs) and antibiotic residuals in diverse environments and their harmful effects on various organisms. DBPs and antibiotics

DOI: 10.1201/9781003120629-6 **43**

found in surface water are generally in a concentration of 0.01 and 1.0 µg/L (Le Page et al., 2017). DBPs are found to inhibit the growth of freshwater algae and duckweeds, thereby restricting the function of photosynthesis and disturbing the interference of chloroplast metabolism at environmental concentrations (Le Page et al., 2017). DBPs have caused oxidative stress and DNA damage and have activated the DNA repair system at environmental concentrations (Ma et al., 2017).

The purview of the recent pandemic and the widespread use of antibiotics have made developing countries the hotspot zone for the emergence of antimicrobial resistance. The self-medication and easy availability of antibiotics without prescriptions have been the bane for the environment. The increasing concentration of micro-pollutants has been the concern in the present pandemic situation. As discussed earlier the increasing use of disinfectants, antibiotics along with the slow and steady increase of secondary micro-plastics in the land and coastline has been a concern for the future. The use of personal protective equipment (PPE) kits, masks, gloves and other paraphernalia had aggravated the current concern for environmental waste generation. The residuals are proposed to modify the behaviour of organisms, bio-accumulates and cause antibiotic resistance in future.

THE INVINCIBLE JOURNEY

The pandemic has demonstrated the fragility of the healthcare systems around the world towards infectious diseases and the ways to handle the situation. The world has been confronted with worst pandemics in history like the "Black Death", which witnessed 50 million deaths in Europe from 1346 to 1353; "Spanish Flu" in 1918 which claimed nearly 40 million people worldwide – with the emergence and re-emergence of a host of viruses such as Zika, Ebola, Severe Acute Respiratory Syndrome (SARS), H1N1 and Middle East Respiratory Syndrome (MERS). The growing population, the rapid urbanisation, unhygienic habitats and overuse of resources have paved the way for future pandemics in developing countries (Jackson, 2016; Drlica and Perlin, 2011).

When the bacteria, viruses, fungi and parasites over the time change and do not respond to the drugs designed to destroy them, then it will make them resistant and more prevalent in the environment. Overuse of medicine in the field of agriculture and human consumption has exacerbated the problem.

Antimicrobial resistant infections could cause 10 million deaths per year by 2050 and the recent pandemic has aggravated and accelerated the resistance. In the present scenario the whole focus is on COVID-19 and we are forgetting the silent emerging concerns coming from the micro-pollutants generated by various sources.

The outbreak of COVID-19 instigated the generation of medical waste globally as a major threat to public health and environment. During treatment and post-treatment a lot of biomedical wastes are generated from healthcare sectors. The wastes are generated from the sample collected from suspected COVID-19 patients, diagnosis, treatment of huge number of patients and disinfection process implemented for infectious and biomedical waste generated from hospitals (Somani et al., 2020; Zambrano-Monserrate et al., 2020).

BIOMEDICAL WASTE GENERATION

Studies reported that Wuhan in China produced more than 240 metric tons of medical wastes every day during COVID-19 outbreak (Saadat et al., 2020). In India, Ahmedabad in Gujarat witnessed 550–600 kg/day to around 1,000 kg/day more waste at the time of the first phase of lockdown (Somani et al., 2020). Many countries in South East Asia similarly witnessed an increasing medical waste of 154–280 million tons during the pandemic. The burgeoning medical wastes become a challenge for the waste management authorities. The existence of SARS-CoV-2 virus over the surface (Van-Doremalen et al., 2020) of the medical waste generated from hospitals (e.g. needles, syringes, bandages, mask, gloves, etc.) has also raised a concern for its proper disposal in order to safeguard from further infection and environmental pollution. Apart from this, the viral infection has increased the use of face masks, hand gloves and other safety equipment, which has added to the concern. Calma (2020) reported that the US trash amount has been increasing due to excessive use of PPE at domestic level. The use of PPE kits has increased worldwide and hence the plastic waste has also soared. Fadare and Okoffo (2020) gave the estimation that in China the daily production of medical mask has increased to 14.8 million till February 2020 and due to lack of proper infectious waste management these wastes (e.g. face masks, hand gloves, etc.) are dumped in open places and also in household wastes (Rahman et al., 2020). Therefore, this unorganised dumping adds to the environment pollution by creating clogs in water ways (Singh et al., 2020; Zambrano-Monserrate et al., 2020).

PLASTICS PREVALENCE

The potential source of microplastic fibres in the environment is found to be through face masks and other plastic protective based equipment (Fadare and Okoffo, 2020). Dioxin and other toxic elements are released into the environment through the use of polypropylene for N95 masks and Tyvek for protective suits, gloves and medical face shields (Singh et al., 2020). The mixing up of organic and household waste with hazardous medical waste increases the risk of disease transmission and exposure of the virus load in waste workers (Ma et al., 2020; Somani et al., 2020; Singh et al., 2020). The overwhelming load of municipal waste (both organic and inorganic) has detrimental effects on environment like air, water and soil pollution (Islam and Bhuiyan, 2016).The increase in shipped package due to pandemic and quarantine policies has led to a spike in the amount of household waste, especially plastic waste (Somani et al., 2020; Zambrano-Monserrate et al., 2020). Waste recycling is an effective way to prevent pollution, save energy and conserve natural resources (Ma et al., 2019); however, given the present pandemic situation and rapid transmission risk of viral infection, most countries have postponed the waste recycling process. The US restricted recycling programmes in many of its cities (nearly 46%) in order to control the spread of infection (Somani et al., 2020). Many of the European countries, United Kingdom and Italy have restricted the infected residents from waste segregation (Zambrano-Monserrate et al., 2020). The disruption in waste management, recovery and recycling has increased the landfills and accumulation of environmental pollutants across the world.

The rapid uses of disinfectants to terminate the SARS-CoV-2 virus have killed the beneficial non-targeted species also and therefore created an ecological imbalance (Islam and Bhuiyan, 2016). There were studies which reported that SARS-CoV-2 virus was also detected in the infected patient's faeces and in municipal wastewater of many countries like India, Sweden, Netherlands and the US (Ahmed et al., 2020; Nghiem et al., 2020; Mallapaty, 2020). The need for strengthening the treatment process of wastewater has been of paramount importance in order to safeguard people from infection. Therefore, many countries like China have enhanced their disinfection process (escalated the use of chlorine) to prevent the spread of SARS-CoV-2 virus through wastewater. The excessive use of chlorine has other repercussions, such as generation of harmful by-products (Zambrano-Monserrate et al., 2020) which will be evident in future health impacts.

BACTERIAL RESISTANCE

According to recent data published in Europe, the antimicrobial resistance (AMR) has gained momentum as a silent tsunami that will show its worse significance with climate change. Apart from the exacerbated storms and rising sea levels, climate change is showing its prominence in the occurrence of AMR. The two challenges of climate change and AMR are intertwined and require a novel approach to be resolved.

The change in temperature caused by climate change has resulted in an increase in thermal adaptations in microbes which have augmented the higher levels of AMR (Fouladkhah et al., 2020). The world needs climate action as the increase in temperature will lead to further spread of infectious diseases and proper antibiotic treatment is required in both human and animal populations.

Every 1° C increase in temperature correlates with an increase in resistance of 1.02 times for methicillin-resistant *Staphylococcus aureus,* 1.03 times for carbapnem-resistant *Klebsiella pneumoniae* and 1.01 times for multidrug-resistant *Escherichia coli.*

Improving the sanitation infrastructure, testing capabilities and improving access to healthcare (WHO, 2021) are the ways in which populations can control AMR. According to Brauner et al. (2016) and Balaban et al. (2019), bacteria transcend into three primary mechanisms to survive and grow in an antibiotic-rich environment. The physiological change leading to the restricted growth is known as tolerance (Handwerger and Tomasz, 1985; Kester and Fortune, 2014). Secondly, when only some of the cells are in slow growing or non-growing state and still able to survive in antibiotics, it is known as persistence (Balaban et al., 2004; Wakamoto et al., 2013). Finally, when the alteration in their genetic modifications helps the bacteria to survive in higher concentration of antibiotics for a longer period, then it is known as resistance. Various environmental factors such as temperature, pH and availability of nutrients play a detrimental role in the mechanism and survival chances of bacteria in the presence of antibiotics.

The presence of antibiotic-resistant bacteria and genes in warm temperatures has been witnessed in wastewater treatment plants across Europe (Parnanen et al., 2019). Further investigation for the presence of antibiotic-resistant bacteria and genes in influent and effluent in twelve wastewater treatment systems situated in seven European nations gave the rationale that temperature plays a pivotal factor in antibiotic resistance when measured with the persistence of resistance and the spread in the environment (Parnanen et al., 2019). There are various ways in which higher temperature can affect resistance, i.e., increase in mutation along with potential for resistance, bacterial persistence and shift of community composition and the rapid dispersal across the spatial scales.

METHODS OF DISINFECTION

Various methods of disinfection in the present scenario of pandemic also add to the woes and it is of paramount importance to treat the wastewater before it is being disposed off in surface water. The wastewater coming from healthcare facilities, household and other public amenities needs to be disinfected in order to treat the microorganisms, and the present prevalent one is COVID-19 (Crini and Lichtfouse, 2019; Li et al., 2020; Wang et al., 2020; Choi et al., 2021). There are various methods of treatment that are used for the same.

Ultraviolet radiation was extensively used from 1910 (Leifels et al., 2019) for disinfection of wastewater and these radiations range from 200 to 400 nm. These radiations are divided into four wave bands on the basis of characteristic wavelengths: ultraviolet A (315 to 400 nm), ultraviolet B (280 to 315 nm), ultraviolet C (200 to 280 nm) and vacuum (100 to 200 nm). The wavelength range of 200 nm and above has the ability to degrade the genetic material of microorganisms, including bacteria and viruses, thereby preventing protein synthesis. Ultraviolet B and C can be used for disinfection of wastewater as they have a good bactericidal effect. This type of disinfection is more economical than chlorine. Advanced processes such as photocatalysis, ozone combined with ultraviolet/hydrogen peroxide, or both ultraviolet and hydrogen peroxide are used for efficient treatment of wastewater.

Chlorine-based disinfectants where hypochlorous acid and hypochlorite ions are used are the most effective way for dealing with viral contamination (Lee et al., 2018). Eminent sources of free available chlorine are sodium hypochlorite, calcium hypochlorite chlorine dioxide, chloramines and elemental chlorine. Hypochlorite being a powerful oxidising agent is potent for oxidising organic pollutants with undissociated hypochlorous acid that acts like a microbial agent (Chatterjee, 2020). Due to the lack of metabolic enzyme system the viruses exhibit more tolerance towards chlorine than bacteria. pH plays a decisive role in inactivation of viruses;

at low pH, the inactivation rate is more compared to that at high pH. pH is the main factor that determines the dissociation of hypochlorous acid to a lesser form such as hypochlorite ion (Zhang et al., 2020).

Hypochlorite is the most common chlorine disinfectant used, which is considered an effective virus disinfectant against SARS-CoV compared to chlorine dioxide (Zhang et al., 2020). Though it is considered as a broad-spectrum microbial disinfectant, its efficacy is lower against viruses at a high pH. It can be used on a small scale for virus disinfection as it is cost-effective, easy to handle and has low residual toxicity.

Chlorine dioxide is used as an alternative for chlorine (Lee et al., 2018). It is adsorbed in proteins (capsomeres) of viruses where it reacts with ribonucleic acid. Chlorine dioxide has more efficacy than chlorine and ozone against viruses.

Hydrogen peroxide is also used for the treatment of wastewater as it is a more economical, and advantageous oxidizing alternative which is effective against yeast, bacteria, spore, fungi and viruses.

Ozone is a reliable and powerful microbial impact used against viruses, bacteria and protozoan (Tizaoui, 2020). Ozone is successful in destroying the viral protein such as capsid protein, protein hydroperoxides and protein hydroxides; therefore, they generate oxidative stress which incapacitates the combating characteristics of virus. It is a substantial disinfectant that enhances biological water quality in less time and concentration.

The residuals or by-products of the disinfectants cause various health problems in the long run and also increase the dissolved organic carbon (DOC) concentrations that affect the treatment given to the wastewater and drinking water. Changes in climatic conditions affect the disinfectants used as the microorganisms tend to get resistant with the change in temperature. The increasing use of disinfectants, antibiotics and micro-pollutants in the form of masks, PPE kits, etc., will have a huge impact in the forthcoming man-made calamities. If we cater to the problems in the present scenario, then it will be easier for us to safeguard the banking from all the future turbulences.

REFERENCES

1. Watkinson, A.J., Murby, E.J. and Costanzo, S.D., 2007. Removal of antibiotics in conventional and advanced wastewater treatment: implications for environmental discharge and wastewater recycling. *Water Res, 41*(18), 4164–4176.
2. Kovalova, L., Siegrist, H., Singer, H., Wittmer, A., and McArdell, C.S., 2012. Hospital wastewater treatment by membrane bioreactor: performance and efficiency for organic micropollutant elimination. *Environ Sci Technol, 46*(3), 1536–1545.
3. Chatterjee, A., 2020. Use of hypochlorite solution as disinfectant during COVID-19 outbreak in India: from the perspective of human health and atmospheric chemistry. *Aerosol Air Quality Res, 20*(7), 1516–1519.
4. Le Page, G., Gunnarsson, L., Snape, J. and Tyler, C.R., 2017. Integrating human and environmental health in antibiotic risk assessment: a critical analysis of protection goals, species sensitivity and antimicrobial resistance. *Environ Int, 109*, 155–169.

5. Ma, L.P., Li, A.D., Yin, X.L. and Zhang, T., 2017. The prevalence of integrons as the carrier of antibiotic resistance genes in natural and man-made environments. *Environ Sci Technol, 51*(10), 5721–5728.

6. Somani, M., Srivastava, A.N., Gummadivalli, S.K. and Sharma, A., 2020. Indirect implications of COVID-19 towards sustainable environment: an investigation in Indian context. *Biores Technol Rep, 11*,100491.

7. Zambrano-Monserrate, M.A., Ruanob, M.A. and Sanchez-Alcalde, L., 2020. Indirect effects of COVID-19 on the environment. *Sci Total Environ, 728*, 138813.

8. Saadat, S., Rawtani, D. and Mustansar, C., 2020. Hussain environmental perspective of COVID-19. *Sci Total Environ, 728*, 138870.

9. Van-Doremalen, N., Bushmaker, T., Morris, D.H., Holbrook, M.G., Gamble, A., Williamson, B.N., et al., 2020. Aerosol and surface stability of SARSCoV-2 as compared with SARS-CoV-1. *N Engl J Med, 382*(16), 1564–1567.

10. Calma, J., 2020. The COVID-19 pandemic is generating tons of medical waste. *The Verge, Mar, 26*, 2020.

11. Fadare, O.O. and Okoffo, E.D., 2020. Covid-19 face masks: a potential source of microplastic fibers in the environment. *Sci Total Environ, 737*, 140279.

12. Rahman, M.M., Bodrud-Doza, M., Griffiths, M.D. and Mamun, M.A., 2020. Biomedical waste amid COVID-19: perspectives from Bangladesh. *The Lancet. Global Health*, 8(10), e1262.

13. Singh, N., Tang, Y. and Ogunseitan, O.A., 2020. Environmentally sustainable management of used personal protective equipment. *Environ Sci Technol*, 54(14), 8500–8502

14. Islam, S.M.D. and Bhuiyan, M.A.H., 2016. Impact scenarios of shrimp farming in coastal region of Bangladesh: an approach of an ecological model for sustainable management. *Aquacult Int, 24*(4), 1163–1190.

15. Ahmed, W., Angel, N., Edson, J., Bibby, K., Bivins, A. and O'Brier J.W., 2020. First confirmed detection of SARS-CoV-2 in untreated wastewater in Australia: a proof of concept for the wastewater surveillance of COVID-19 in the community. *Sci Total Environ, 728*, 138764.

16. Nghiem, L.D., Morgan, B., Donner, E. and Short, M.D., 2020. The COVID-19 pandemic: considerations for the waste and wastewater services sector. *Case Stud Chem Environ Eng, 1*, 100006.

17. Mallapaty, S., 2020. How sewage could reveal true scale of coronavirus outbreak. *Nature,580*, 176–177.

18. Jackson, M., 2016. *The Routledge history of disease*. New York, NY: Routledge.

19. Drlica, K. and Perlin, D., 2011. *Antibiotic resistance: Understanding and responding to an emerging crisis*. Upper Saddle River, NJ: FT Press.

20. World Health Organization, 2021. Case study: health care without harm Europe SAICM 2.0 project: promoting safer disinfectants in the health-care sector: over 80 health providers from 19 countries..

21. Fouladkhah, A., Thompson, B. and Camp, J., 2020. The threat of antibiotic resistance in changing climate. *Microorganisms*, 8(5), 748.

22. Balaban, N.Q., Helaine, S., Lewis, K., Ackermann, M., Aldridge, B., Andersson, D.I., et al., 2019. Definitions and guidelines for research on antibiotic persistence. *Nat Rev Microbiol, 17*, 441–448.

23. Brauner, A., Fridman, O., Gefen, O. and Balaban, N.Q., 2016. Distinguishing between resistance, tolerance and persistence to antibiotic treatment. *Nat Rev Microbiol, 14*, 320–330.

24. Handwerger, S. and Tomasz, A., 1985. Antibiotic tolerance among clinical isolates of bacteria. *Annu Rev Pharmacol Toxicol, 25,* 349–380.
25. Kester, J.C. and Fortune, S.M., 2014. Persisters and beyond: mechanisms of phenotypic drug resistance and drug tolerance in bacteria. *Crit Rev Biochem Mol Biol, 49,* 91–101.
26. Balaban, N.Q., Merrin, J., Chait, R., Kowalik, L. and Leibler, S., 2004. Bacterial persistence as a phenotypic switch. *Science., 305,* 1622–1625.
27. Wakamoto, Y., Dhar, N., Chait, R., Schneider, K., Signorino-Gelo, F., Leibler, S. et al., 2013. Dynamic persistence of antibiotic-stressed Mycobacteria. *Science., 339,* 91–95.
28. Parnanen, K.M.M., Narciso-da-Rocha, C., Kneis, D., Berendonk, T.U., Cacace, D., Do, T.T., et al., 2019. Antibiotic resistance in European wastewater treatment plants mirrors the pattern of clinical antibiotic resistance prevalence. *Sci Adv, 5,* eaau9124.
29. Crini, G. and Lichtfouse, E. 2019. Advantages and disadvantages of techniques used for wastewater treatment. *Environ Chem Lett, 17,* 145–155.
30. Li, C., Yang, J., Zhang, L., Li, S., Yuan, Y., Xiao, X., et al., 2020. Carbon-based membrane materials and applications in water and wastewater treatment: a review. *Environ Chem Lett, 19*(2), 1457–1475.
31. Wang, X., Han. J. and Lichtfouse, E., 2020. Unprotected mothers and infants breastfeeding in public amenities during the COVID-19 pandemic. *Environ Chem Lett, 18,* 1447–1450.
32. Choi, H., Chatterjee, P., Lichtfouse, E., Martel, J.A., Hwang, M., Jinadatha, C. et al., 2021. Classical and alternative disinfection strategies to control the COVID-19 virus in healthcare facilities: a review. *Environ Chem Lett, 19,* 1945–1951.
33. Leifels, M., Shoults, D., Wiedemeyer, A., Ashbolt, N.J., Sozzi, E., Hagemeier, et al., 2019. Capsid Integrity qPCR—an azo-dye based and culture-independent approach to estimate adenovirus infectivity after disinfection and in the aquatic environment. *Water, 11,* 1196.
34. Lee, H.W., Lee, H.M., Yoon, S.R., Kim, S.H. and Ha, J.H., 2018. Pretreatment with propidium monoazide/sodium lauroyl sarcosinate improves discrimination of infectious waterborne virus by RT-qPCR combined with magnetic separation. *Environ Pollut, 233,* 306–314.
35. Zhang, D., Ling, H., Huang, X., Li, J., Li, W., Yi, C., et al., 2020. Potential spreading risks and disinfection challenges of medical wastewater by the presence of severe acute respiratory syndrome coronavirus 2 (SARS-CoV-2) viral RNA in septic tanks of Fangcang Hospital. *Sci Total Environ, 741,* 140445.
36. Tizaoui, C., 2020. Ozone: a potential oxidant for COVID-19 virus (SARS-CoV-2). *Ozone Sci Eng, 42,* 378–385.
37. Ma, Y., Lin, X., Wu, A., Huang, Q., Li, X. and Yan, J., 2020. Suggested guidelines for emergency treatment of medical waste during COVID-19: Chinese experience. *Waste Disposal & Sustainable Energy, 2,* 81–84.
38. Ma, B., Li, X., Jiang, Z and Jiang, J., 2019. Recycle more, waste more? When recycling efforts increase resource consumption. *Journal of Cleaner Production, 206,* 870–877.

7 Healthcare Directives Abolishing Risks

CHANGES LEADING TO UNCERTAINTIES

The rapid and enormous environmental changes that occurred in the previous decade have borne the seed of the current pandemic (COVID-19). Many predictions were made pertaining to the spread of zoonotic diseases (Wu et al., 2016; Baylis, 2017), depleting food security (Wheeler and von Braun, 2013) and vulnerability in human immune system (Swaminathan et al., 2014) with the changing environmental factors. With the change in climate there would be prevalence of newer emerging diseases.

The climatic aberrations and the increasing temperature will have an impact on the seasonal peak and time window of the future, with potential epidemic viral infections such as zoonotic viral diseases (Liu-Helmersson et al., 2016).

The global mean atmospheric carbon dioxide (CO_2) level has increased over the last few decades compared to the past 800,000 years (Lüthi et al., 2008). There are complicated and multidimensional links between a human-induced climate change and global health risks (McMichael et al., 2008; Butler, 2018).

The climatic change contributes to the expansion of geographical distribution of infectious diseases by affecting the pathogens, vectors, hosts, and/or their living environment (Wu et al., 2016). The climatic change has a potent impact on the replication, development and transmission rate of pathogens (Ruiz et al., 2010; Ruiz-Moreno et al., 2012; Altizer et al., 2013). In vitro the viral replications are affected by the change in temperature and in vivo they affect the frequency of transmission in various models (Lowen et al., 2007; Foxman et al., 2016; Moriyama and Ichinohe, 2019).

Global warming played a vital role in the viral epidemics such as the West Nile virus where it affects the rate of evolution of virus as well as intensifying and preserving the human infection (Paz, 2015).

Bats are the known natural conservatory for various viruses, such as the severe acute respiratory syndrome and Middle East respiratory syndrome viruses, that caused the previous coronavirus outbreaks (O'Shea et al., 2014). In the present pandemic situation, these mammals are suspected to be the natural hosts for severe

DOI: 10.1201/9781003120629-7

acute respiratory syndrome-coronavirus-2 (SARS-COV-2), the causative agent of COVID-19 (Sun et al., 2020). The increase in the novel coronavirus within bat communities in the past few years is predicted to be due to the impact of climate change on their geographic locations and habitat suitability (Lorentzen et al., 2020). According to Prada et al. (2019), there is a high prevalence of coronavirus spreading and increase in viral load in Western Australia due to the distributional shifting of the bat populations in order to avoid the implications of climate change that abridged their habitats.

Climatic change has also impacted the timing of the migratory birds and therefore they have escalated the spread of viral infection prevalence and spread (Brown and Rohani, 2012).

The implications of stressful events, such as climate change and destruction of habitat, will lead to an alteration of the immune tolerance and significant increase of the viral replication. The viral load of the persistently infected bats will also contribute to the spread of the virus (Chionh et al., 2019; Subudhi et al., 2019). The marked depletion in the interval of viral outbreak in bats was observed to be caused by several factors such as climate change, food insecurity and population growth (Plowright et al., 2015; Wang and Anderson, 2019).

Due to the change in climate there is a change in temperature, atmospheric CO_2 and cloud cover. The factors cited have a direct effect on the growth of plants and trees. The climatic changes altering the environmental factors have a direct effect on natural habitats and ecosystems. Subtle adjustments have greater impacts on species thriving within an ecosystem. The present pandemic (COVID-19) is invariably linked with climate change(Gorji and Gorgji, 2021). As per the World Health Organization (WHO), there has always been a link between environmental conditions and epidemic diseases. These links in future might compel the policymakers to take into consideration the impact of climate change and also chalk out the strategy along with economic ways to prevent such environmental damages.

INNOVATION FOR TRANSITION

Innovation in healthcare has a link with respect to global climate change as there are significant increases in the concentration of greenhouse gases (GHGs) in the atmosphere like CO_2, methane (CH_4) and nitrous oxide (N_2O), which is further concentrated by the addition of industrial GHGs such as hydrofluorocarbons (HFCs), perfluorocarbons (PFCs) and sulphur hexafluoride (SF_6).

GHGs instigate climate change by tapping heat in the atmosphere, and therefore it raises the average temperature of the planet. There are significant alterations also registered in the patterns and intensity of precipitation apart from the flow of air and ocean currents across the globe. Human activities leading to the GHG emissions are the main source of increased GHGs in the atmosphere (IPCC, 2007).

Research data has made it possible to understand and analyse the link between human activities, GHG emissions, increase in atmospheric concentration and the consequent changes in global temperature (Table 7.1). Pertaining to present

TABLE 7.1
The Major Greenhouse Gases and Common Sources of Emissions

Symbols	GHGs	Emissions
CO_2	Carbon dioxide	Fossil fuel combustion, forest clearance for pasture land and cement production, etc.
CH_4	Methane	Landfills, production and distribution of natural gases and petroleum, fermentation from the digestive system of livestock, rice cultivation, fossil fuel combustion, etc.
N_2O	Nitrous Oxide	Fossil fuel combustion, fertilisers, nylon production, manure, etc.
HFCs	Hydroflurocarbons	Refrigeration gases, aluminium smelting, semiconductor manufacturing, etc.
PFCs	Perfluorocarbons	Aluminium production, semiconductor industry, etc.
SF_6	Sulphur hexafluoride	Electrical transmissions and distribution systems, circuit breakers, magnesium production, etc.

Source: IPCC (2007).

situations, policymakers have augmented the notion that no more than a 2°C increase in global temperature as per the climate policy goal is preferred for soliciting the dangerous impacts. It would require actions to stabilise the atmospheric GHG concentrations at a level minimally greater than current levels. According to the Intergovernmental Panel on Climate Change (IPCC, 2007), it would require a reduction in annual global GHG emissions of 50% to 80% below by 2050.

Presently 85% of the world's energy is provided by fossil fuels where half of the same is in the form of oil, coal and natural gas. CO_2 emitted from the combustion of those fuels like power plants and automobiles is the key source of GHG released. The major crux lies in restricting the dangerous climate change to a sustainable low-carbon energy system.

A huge paradigm shift will help to elevate large reductions in global GHG emissions. There are four major strategies to transform the energy system of a nation:

- Curtailing the demands for energy in all major sectors of the economy (buildings, transportation and industry), thus reducing the requirement for fossil fuels;
- Upgrading the efficiency of energy utilisation for reduced requirement of fossil fuel to cater the "end use" energy demands, resulting in lower CO_2 emissions;
- Changing of high-carbon fossil fuels such as coal and oil with lower-carbon or zero-carbon alternatives like natural gas, nuclear and renewable energy sources such as biomass, wind and solar;

- Capturing and sequestering the CO_2 emitted by the combustion of fossil fuels to prevent its release into the atmosphere.

The major challenges in the healthcare system can be avoided by incorporating scientific strategies. The recent pandemic (COVID-19) gave a major challenge to the health sector and the strategies implemented by various nations, which further divided them into two sections. One section strictly adopted the benefits of technological facilities to safeguard the nation and the other section didn't follow any basic norms, leading to a lot of vulnerability towards the disease. The two types of steps taken to control the pandemic around the world are:

1. Mitigation method – This is also known as herd immunity approach which was planned to manage the rate of infection in such a way as to avoid overwhelming the healthcare system and build up enough recovered as well as robust immune people in the population to ultimately interrupt virus transmission.
2. Suppression strategy – This strategy involves physical distancing and travel restrictions (lockdown) to suppress the virus transmission.

Apart from that the elimination strategy was also implemented, which is a known method for controlling infectious diseases as given by epidemiologists. In this method the reduction of the incidence of a disease was targeted to be zero in a defined geographical province. The ultimate goal of this approach is the elimination criteria for highly infectious diseases. The eradication out here means the reduction of diseases to zero at the global level, at least outside laboratories.

The approach towards controlling the pandemic was a bit different in some South East Asian countries such as Vietnam, where stringent control measures including quarantine, control tracing, border controls, school closures and traffic restrictions were adhered when COVID-19 cases were low. This approach was also implemented in Taiwan, Hongkong and South Korea.

The rapid testing method along with implementation of artificial intelligence technologies, wastewater surveillance (Randazzo et al., 2020) and Internet of Things (IoT) applications will generate data that will help the public health officials to understand the extent of COVID-19 infection in communities. These healthcare directives will help a nation to understand the severity of the infection rate and take judicious approach on time.

The other area that is often neglected is the management of infectious waste. The COVID-19 waste strategy should be followed with proper planning as the workforce is less with the rapid rise in the pandemic and the waste generation is humungous. The development of a comprehensive disposal mode signifies the combination of centralised disposal and on-site emergency disposal of medical waste. The treatment facilities implemented for medical waste should be made more automated and depend on the technology based on Internet of Things (IoT) where real-time approach was taken to track and control the disposal method.

These automated processes and the use of minimum workers for the infectious waste were also covered as it includes sensing the equipment information, location system, scanning devices and video surveillance along with Internet access with each device.

With a scarcity of workforce the amount of biomedical waste if left untreated will decay and help in generating a lot of GHGs that will eventually lead to crisis. Due to climate change, factors such as temperature, atmospheric CO_2 and cloud cover are evolving.

In order to control the risk in the healthcare sector, the present pandemic situation is a big challenge but with clear strategic moves and planned directives these can be avoided. The basic approach of invention is driven in a large part by research and development (R&D) that is for both basic and applied research; along with this the innovation plays colloquially to describe the overall technological approach.

REFERENCES

1. Gorji, S. and Gorji, A., 2021. COVID-19 pandemic: the possible influence of the long-term ignorance about climate change. *Environ Sci Pollut Res*, *28*(13), 15575–15579.
2. Wu, X., Lu, Y., Zhou, S., Chen, L. and Xu, B., 2016. Impact of climate change on human infectious diseases: empirical evidence and human adaptation. *Environ Int*, *86*, 14–23.
3. Baylis, M., 2017. Potential impact of climate change on emerging vector-borne and other infections in the UK. *Environ Health*, *16*(Suppl 1), 112.
4. Wheeler, T. and von Braun, J., 2013. Climate change impacts on global food security. *Science*, *341*(6145), 508–513.
5. Swaminathan, A., Lucas, R.M., Harley, D. and McMichael, A.J., 2014. Will global climate change alter fundamental human immune reactivity: implications for child health? *Children (Basel)*, *1*(3), 403–423.
6. Liu-Helmersson, J., Quam, M., Wilder-Smith, A., Stenlund, H., Ebi, K., Massad, E. et al. 2016. Climate change and Aedes vectors: 21st century projections for dengue transmission in Europe. *EBioMedicine*, *7*, 267–277.
7. Lüthi, D., Le Floch, M., Bereiter, B. Blunier, T., Barnola, J-M., Siegenthaler, U., et al., 2008. High-resolution carbon dioxide concentration record 650,000-800,000 years before present. *Nature*, *453*(7193), 379–382.
8. McMichael, A.J., Friel, S., Nyong, A. and Corvalan, C., 2008. Global environmental change and health: impacts, inequalities, and the health sector. *BMJ*, *336*(7637), 191–194.
9. Butler, C.D., 2018. Climate change, health and existential risks to civilization: a comprehensive review (1989-2013). *Int J Environ Res Public Health*, *15*(10), 2266.
10. Ruiz, M.O., Chaves, L.F., Hamer, G.L., Sun, T., Brown, W.M., Walker, E.D., et al., 2010. Local impact of temperature and precipitation on West Nile virus infection in Culex species mosquitoes in northeast Illinois, USA. *Parasit Vectors*, *3*(1), 19.
11. Ruiz-Moreno, D., Vargas, I.S., Olson, K.E. and Harrington, L.C., 2012. Modeling dynamic introduction of Chikungunya virus in the United States. *PLoS Negl Trop Dis*, *6*(11), e1918.

12. Altizer, S., Ostfeld, R.S., Johnson, P.T., Kutz, S. and Harvell, C.D., 2013. Climate change and infectious diseases: from evidence to a predictive framework. *Science, 341*(6145), 514–519.

13. Lowen, A.C., Mubareka, S., Steel, J. and Palese, P., 2007. Influenza virus transmission is dependent on relative humidity and temperature. *PLoS Pathog, 3,* 1470–1476.

14. Foxman, E.F., Storer, J.A., Vanaja, K., Levchenko, A. and Iwasaki, A., 2016. Two interferon-independent double-stranded RNA-induced host defense strategies suppress the common cold virus at warm temperature. *Proc Natl Acad Sci USA, 113,* 8496–8501.

15. Moriyama, M. and Ichinohe, T., 2019. High ambient temperature dampens adaptive immune responses to influenza A virus infection. *Proc Natl Acad Sci USA, 116*(8), 3118–3125.

16. Paz, S., 2015. Climate change impacts on West Nile virus transmission in a global context. *Philos Trans R Soc Lond B Biol Sci., 370*(1665), 20130561.

17. O'Shea, T.J., Cryan, P.M., Cunningham, A.A., Fooks, A.R., Hayman, D.T.S., Luis, A.D., et al., 2014. Bat flight and zoonotic viruses. *Emerg Infect Dis, 20*(5), 741–745.

18. Sun, Z., Thilakavathy, K., Kumar, S.S., He, G. and Liu, S.V., 2020. Potential factors influencing repeated SARS outbreaks in China. *Int J Environ Res Public Health, 17*(5), 1633.

19. Lorentzen, H.F., Benfield, T., Stisen, S. and Rahbek, C., 2020. COVID-19 is possibly a consequence of the anthropogenic biodiversity crisis and climate changes. *Dan Med J, 67*(5), A205025.

20. Prada, D., Boyd, V., Baker, M.L., O'Dea, M. and Jackson, B. 2019. Viral diversity of microbats within the south west botanical province of western Australia. *Viruses, 11*(12), 1157.

21. Chionh, Y.T., Cui, J., Koh, J., Mendenhall, I.H., Ng, J.H.J., Low, D., et al., 2019. High basal heat-shock protein expression in bats confers resistance to cellular heat/oxidative stress. *Cell Stress Chaperon, 24*(4), 835–849.

22. Brown, V.L. and Rohani, P., 2012. The consequences of climate change at an avian influenza "hotspot". *Biol Lett, 8*(6), 1036–1039.

23. Subudhi, S., Rapin, N. and Misra, V. 2019. Immune system modulation and viral persistence in bats: understanding viral spillover. *Viruses, 11*(2), 192.

24. Plowright, R.K., Eby, P., Hudson, P.J., Smith, I.L., Westcott, D., Bryden, W.L., et al., H., 2015. Ecological dynamics of emerging bat virus spillover. *Proc Biol Sci, 282*(1798), 20142124.

25. Wang, L.F. and Anderson, D.E., 2019. Viruses in bats and potential spillover to animals and humans. *Curr Opin Virol, 34*:79–89.

26. Randazzo, W., Cuevas-Ferrando, E., Sanjuán, R., Domingo-Calap, P. and Sánchez, G., 2020. Metropolitan wastewater analysis for COVID-19 epidemiological surveillance. *Int J Hyg Environ Health, 230,* 113621.

27. IPCC. 2007. *Climate change 2007: synthesis report. Summary for Policymakers.* Geneva: IPCC.

8 Preventive, Prerogative and Protective Measures

PROACTIVE APPROACH

The proactive approach of a nation always safeguards it from the dire consequences of climate change and its impact on people. The emerging climate change has paved the way for so many hurdles in the environment in the form of epidemiological triad. The humans (the host), pathogens (the agent) and vectors (the environment) are the three triad. With the emerging climate change there is a change in the temperature, humidity, availability of water and food.

In the due course of time, climate change is framed in varied ways. It is framed as an "environmental problem" that can be understood through the anthropogenic impacts on the natural environment, ecosystems and particular species, e.g., polar bears, coral reefs and tropical rainforests.

Another common frame that is portrayed as Pandora's box has a long list of potentially disastrous impacts, e.g., sea level rise, drought, floods, heat waves, infectious diseases, famines and depletion of water level (Nisbet, 2009). If the issue of climate change is not addressed in the present scenario, it will induce degradation and undermine the future productivity, leading to economic crisis in a nation. The outbreak of the COVID-19 pandemic at the beginning of 2020 intensified the crisis, causing a drastic decline in aggregate demand and output.

The three future scenarios for the marine environment that are discussed in the text are commercial fisheries and trade, coastal communities and species conservation.

Climate change has made so many lasting impacts on the environment. The three scenarios as depicted in Figure 8.1 give us a glimpse of the impact on the marine environment. As per the report of Aschey (2018), climate change has a deep impact on the harvest of fish and the migration of fish has hampered the population dependent on fisheries. According to IPCC (2014), climate change has influenced the migration of species and the jurisdictional issue will pose a problem for the people whose means of survival depend on pisciculture. The variation in fish population and the ecosystem from the impact of climate change directly affect the fishing communities and their operations. Scientists have recognized the

DOI: 10.1201/9781003120629-8

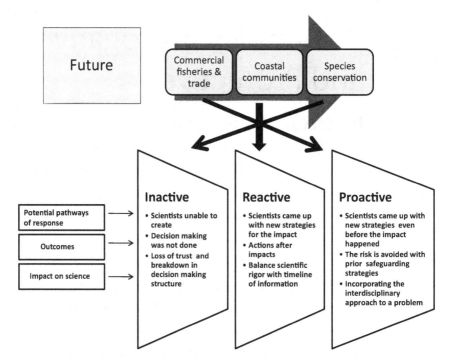

FIGURE 8.1   Climate change presents opportunity for a spectrum of pathways of response by scientists and the pathways (inactive, reactive, and proactive) have different outcomes and impact on science.

problems faced by the communities and have come up with innovative remedies. Links between physical changes and warming are visible in hydrology and ocean-ography, and the ecological behaviour aspect change in food webs and operation of the fishing sectors as per the economics are complex problems and need due attention to resolve.

The activity of the wave and the impact of the storm will have the following reduction impacts: the number of fishing days, ability to migrate for the fish and seasonal availability of fish (Hobday and Poloczanska, 2010).

The two main impacts on the fisheries were categorised into micro- and macro-economic analysis:

- Within the sector: will have impact on incomes, assets, livelihood of indi-vidual fishers, processors, fish farmers and the communities;
- Sector contribution to national economies: impacts on revenues, exports, per capita fish supply and the contribution to employment and national level gross domestic product (GDP).

THE CHANGE IN CLIMATE HAS IMPLICATIONS FOR FISHERS, COMMUNITIES AND NATIONAL ECONOMIES

The following are the changes associated with fisheries, communities and national economies for the two sectors.

Within the Sector

1. Innovative fishing methods;
2. Systematic change in aquaculture production;
3. Alterations in marketing chains;
4. The division of balance between fishing, aquaculture and other livelihood activities;
5. To invest in community or participatory resource management.

Sector Contribution to National Economies

1. Government to allocate rent from fisheries;
2. Change observed for food and nutritional security;
3. Employment augmented for safety nets, social and cultural role of fisheries;
4. Adaptation cost.

Alterations along with an innovative approach will help establish a proper measure for safeguarding the aquatic ecosystem amidst the climatic change.

THE INTERRELATIONSHIP BETWEEN THE HUMAN AND EARTH SYSTEM

There is a coherent relationship between the climatic conditions prevailing in the human and earth system. The earth system is dependent on climate change, process drivers, impacts and vulnerability, whereas the human systems depend on socio-economic development as depicted in Figure 8.2. The prerogative measures in the healthcare system in the form of personal protective equipment (PPE) kit, masks, etc., have safeguarded us from the impact of pandemic.

The concentrations of pollutants need to be reduced and it will help in building a strong ecosystem, water resources, food security, settlement and security and implication on human health.

The human system depends on the amount of socio-economic development which further depends on mitigation and adaptations. These preventive and protective measures will further promulgate a uniformed balance in both the systems.

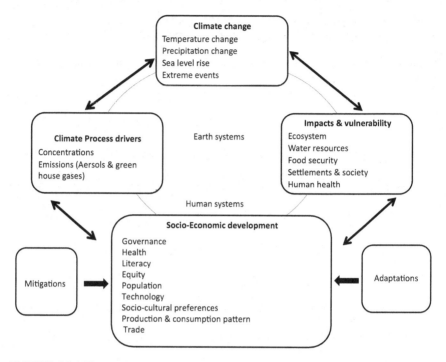

FIGURE 8.2 The earth systems: climate change, climate process drivers, impacts and vulnerability. Human systems: socio-economic development (climate process drivers, impacts and vulnerability).

REFERENCES

1. Nisbet, M.C., 2009. Communicating climate change: why frames matter for public engagement. *Environ Sci Policy for Sustain Dev*, *51*(2), 12–23.
2. Asche, F., 2018. Rome 2018 report entitled "Impacts of climate change on the production and trade of fish and fishery products". The State of Agricultural Commodity Markets (SOCO) 2018 - Background Paper. Rome: FAO. 44 pp.
3. Hobday, A.J. and Poloczanska, E.S., 2010. Marine fisheries and aquaculture. In: Stokes C. and Howden M. (eds.), *Adapting agriculture to climate change: preparing australian agriculture, forestry and fisheries for the future* (pp. 205–228). Collingwood: CSIRO Publishing.
4. IPCC., 2014. *Climate change*. Geneva: IPCC.

9 Demographic Dimensions

POPULATION EXPLOSION

The increased population has a bigger impact on the environment as lots of natural resources are utilised and the generation of greenhouse gases has soared up.

With the increasing population there is also a demand for food. To cater to the demand of the population intense farming is being done. Rapid farming has its own impact; to go ahead with farming there is a requirement of enough farmlands and hence the need for deforestation (Kissinger et al., 2012). Agriculture is responsible for 80% of the deforestation. To gain good yield there is an extensive use of pesticides and chemical fertilisers. These practices lead to soil erosion and soil depletion. The agricultural run-off with its rich nutrient content often finds its way in the water bodies and therefore increases the chances of eutrophication (Anderson et al., 2002; Heisler et al., 2008; Gilbert et al., 2005, 2014, 2018). This phenomenon depletes the oxygen level in water and therefore results in significant negative impacts on marine life. The need for land leads to deforestation which in turn reduces the ability to capture carbon dioxide (CO_2), therefore accelerating the greenhouse gas problem. The clearing of forest is strongly associated with the loss of indigenous flora and fauna. And agriculture is credited for the 80% of global deforestation. The other 20% is shared by water logging, firewood collection and fodder for raising and breeding livestock. China's demand for animal feed is one of the major reasons for man-made forest fires in the Amazon. Northern and Southern US soybeans are raised in the Amazon and are exported in exchange for the pork meat produced in China. There was a 30% upsurge in meat production in China in the last few decades (Sheldon, 2019).

The increase in agriculture demand also leads to extensive use of chemical fertilisers along with pesticides in order to have rapid and good yield to sustain the growing population. The chemical fertilisers when used extensively make the land more unsuitable for irrigation, leading to the problems of water logging and nutrients run-off in the nearby aquatic land, hence promulgating eutrophication.

The dense growth of plants during the process of eutrophication consumes oxygen, resulting in the death of aquatic animals. The other major sources of

DOI: 10.1201/9781003120629-9

eutrophication are the industries and sewage disposal which have also increased due to the growth of population.

Our earth consists of three-fourths of water out of which only 2.5% is fresh-water; therefore, the availability of freshwater is scarce as only a very small fraction of 2.5% is unpolluted drinking water. As per the study conducted by United Nations University 2021, the increasing population creates a higher water stress on freshwater supplies. "Water stressed" is defined as a situation where there is a demand exceeding the supply of suitable water. According to a study, around 40% of the world's population endures water scarcity and it is projected to sky-rocket by 2030 since the demand for water will increase by 50% (U.S. Bureau of Reclamation, 2021). The increasing growth of population in near future will exhaust all the natural resources.

The growing population and climate change will go in the same direction as the use of fossil fuel will support the industrialised societies. The more the population, greater will be the demand for oil, coal, gas and other energy sources, which need to be excavated from the ground, leading to spewing of CO_2 into the atmosphere. Apart from this, the rapid growth of greenhouse gases will also add to the above causes.

ECONOMIC ESCALATION THROUGH INDUSTRY

The consumption of fossil fuel is higher in developed countries and developing countries are invariably at par with the developed nations because they have to escalate their economy through industrialisation (Acton et al., 2021).

Rapid deforestation is also a reason for greenhouse gas emissions. In a study conducted by the Food and Agriculture Organization of the United Nations, it was revealed that forests store more than twice the amount of CO_2 available in the atmosphere. As these forests are cleared and burnt, CO_2 released into the atmosphere accounts for a whopping 12% of the total greenhouse gas production (FAO, 2021).

REFERENCES

1. Kissinger, G.M., Herold, M. and De Sy, V., 2012., *Drivers of deforestation and forest degradation: a synthesis report for REDD+ policymakers* Lexeme Consulting.
2. Anderson, D.M., Gilbert, P.M. and Burkholder, J.M., 2002. Harmful algal blooms and eutrophication: nutrient sources, composition, and consequences. *Estuaries*, 25(4), 704–726.
3. Gilbert, P.M., Berdalet, E., Burford, M.A., Pitcher, G.C. and Zhou, M., 2018. Introduction to the global ecology and oceanography of harmful algal blooms (GEOHAB) synthesis. In *Global ecology and oceanography of harmful algal blooms* (pp. 3–7). Cham: Springer.
4. Heisler, J., Gilbert, P.M., Burkholder, J.M., Anderson, D.M., Cochlan, W., Dennison, W.C., et al., 2008. Eutrophication and harmful algal blooms: a scientific consensus. *Harmful Algae, 8*(1), 3–13.

5. Sheldon, I., 2019. Why China's soybean tariffs matter. *Salon*, April 8, 2018.
6. U.S. Bureau of Reclamation., 2021. Water facts - worldwide water supply. Accessed Feb. 5, 2021.
7. United Nations University., 2021. Global water crisis: the facts. Page 3. Accessed Feb. 5, 2021.
8. Food and Agriculture Organization of the United Nations., 2021. Forests and climate change. Accessed Feb. 5, 2021.
9. Gilbert, P.M., Anderson, D.M., Gentien, P., Granéli, E. and Sellner, K.G., 2005. *The global, complex phenomena of harmful algal blooms. Oceanography, 18*(2), 136–147.
10. Gilbert, P., 2014. Harmful Algal Blooms in Asia: an insidious and escalating water pollution phenomenon with effects on ecological and human health. *ASIANetwork Exchange J Asian Stud Lib Arts, 21*(1), 52–68.

10 Assiduous Economic Growth

PERTURBATIONS IN ECONOMY

The grim situation due to the pandemic and the continuous economic downturn has escalated to such a state where people are confined to stay within their dwellings and still monetary crisis is manoeuvring. The healthcare issues have massively been affected and the second wave of COVID-19 created a havoc in most countries, especially India, where millions of people died and there was a huge depreciation in the economic status of the country. The constant stagflation due to recession has aggravated the situation.

Economic perturbations were more severe when the black market in healthcare for medicines, hospital bed reservation and oxygen cylinders were demanding sky-rocketing price for these facilities. Those who can afford were also not saved as the situation was out of control. For the past one year, i.e. 2000, nothing has been done to safeguard the future in healthcare.

The healthcare system around the world was staggering as defined in the review carried out with publicly available information on the economic interventions countries have put in place to ameliorate the impact of COVID-19 (Danielli et al., 2021).

THE FISCAL DISCREPANCY

The financial pressure on businesses and employees was mitigated by economic measures when the countries went for lockdown. Table 10.1 provides us the gross domestic product (GDP) expenditure and financial interventions faced by some of the developed and developing countries (Danielli et al., 2021).

Developing countries have limited resources and in order to sustain in the ever-increasing global market they need to curtail or mitigate the available resources. While the incentive to free ride on other countries' costly climate mitigation is obvious, free riding between countries is less evident for COVID-19 (Fuentes et al., 2020). In the case of COVID-19, the extent to which one affected country can benefit from coping policies undertaken in another country seems limited. One

DOI: 10.1201/9781003120629-10

TABLE 10.1
Summary of Countries' GDP Expenditure and Fiscal Interventions

Country	% GDP equivalent fiscal response	Financial interventions
China	2.50	1. Income support measures for individuals and households excluding tax and contribution changes 2. Tax and contribution policy changes 3. Deferral of taxes and social security contributions 4. Loan guarantees by the state 5. Medical equipment and pharmaceuticals related to the new coronavirus are exempt from registration fees
New Zealand	4.00	1. Income support measures for individuals and households 2. An increase in the provisional tax threshold and the reintroduction of depreciation charges for commercial buildings 3. NZD 600 million will be spent on support for the aviation industry 4. The Business Finance Guarantee Scheme 5. Six-month interest and principal payment holiday for mortgage holders and SMEs who have lost income
Australia	10.50	1. Income support measures for individuals and households 2. Movement restrictions and self-isolation for tourists 3. Increasing the threshold for assets eligible for instant tax write-off 4. Firms that are affected by the pandemic to defer payment of tax liabilities for up to 4 months 5. The government will guarantee 50% of new loans issued by eligible lenders to SMEs
USA	11.00	1. Income support measures for individuals and households excluding tax and contribution changes 2. Tax reductions and deferrals 3. Public sector subsidies to businesses 4. Public sector loans or capital injections to businesses
South Korea	11.40	1. Income support measures for individuals and households 2. Relief checks to households in the bottom 70% income bracket 3. Introduction of temporary special tax reduction for SMEs located in corona-related special disaster areas 4. VAT reduction and exemptions 5. 30% contribution rate deduction for social security for small business and low-income households 6. Public sector loans or capital injections

TABLE 10.1 (Continued)
Summary of Countries' GDP Expenditure and Fiscal Interventions

Country	% GDP equivalent fiscal response	Financial interventions
Spain	12.00	1. Obligatory shutdown of economic activities 2. Income support measures for individuals and households 3. Specific programme for victims of gender violence, homeless people and others who are especially vulnerable 4. Exemption of social security contributions by impacted firms 5. Reduction of VAT applicable to the supply of medical equipment from national producers to public entities 6. Deferral of taxes and social security contributions 7. Public sector loans or capital injections to businesses 8. Loan guarantees by the state
Sweden	14.90	1. Income support measures for individuals and households 2. Removal of the income ceiling for student aid 3. Tax and contribution policy changes 4. Public sector subsidies to businesses 5. Government covering 50% of the rental reduction up to 50% of the fixed rent 6. Deferral of taxes and social security contributions 7. Public sector loans or capital injections to businesses 8. Loan guarantees by the state
France	19.00	1. Employment protection: furlough, short-time working, temporary leave 2. Shutdown of economic activities (easing initiated) 3. Suspension/postponement on payments: tax, VAT, business rates, etc. 4. Direct compensation to businesses and/or consumers for damages caused by COVID-19 5. Aid to support particular industries, e.g., airports 6. State guarantees on loans taken by businesses: SMEs, mid-caps, etc., from banks 7. Credit insurance
Japan	21.70	1. Income support measures for individuals and households 2. Provides SMEs and large corporations a financial support 3. Public sector subsidies to businesses 4. Deferral of taxes and social security contributions 5. Public sector loans or capital injections 6. Loan guarantees by the state

(continued)

TABLE 10.1 (Continued)
Summary of Countries' GDP Expenditure and Fiscal Interventions

Country	% GDP equivalent fiscal response	Financial interventions
UK	22.10	1. Obligatory shutdown of economic activities (some easing in place) 2. Income support measures for individuals and households excluding tax and contribution changes (now extended) 3. Deferral of income tax and VAT payments 4. In England, 100% relief of business rates on property for all properties in retail, hospitality or leisure 5. Universal Credit (UC) for self-employed 6. Public sector loans or capital injections to businesses 7. Loan guarantees by the state benefiting private borrowers 8. Welfare support
Germany	27.00	1. Employment protection: furlough, short-time working, temporary leave 2. State guarantees on loans taken by businesses: SMEs, mid-caps, etc., from banks 3. Subsidies on loan/interests 4. Direct grants 5. Credit insurance 6. Obligatory shutdown of economic activities (easing has initiated) 7. Child allowance (Kinderzuschlag) 8. Reduced VAT rate of 7% will be applicable to restaurants 9. Public sector subsidies to businesses 10. Cover 100% of social security contributions for lost hours of short-time workers
Italy	50.00	1. Income support measures for individuals and households excluding tax and contribution changes 2. Moratorium on debt payments, including mortgages 3. 60% tax credit on commercial rents 4. Fund to provide fee-free guarantee for SMEs loans 5. Deferral of taxes and social security contributions 6. Loan guarantees by the state 7. One-year suspension in the repayment of real estate mortgages

SME: Small and medium-sized enterprises

Source: Danielli et al. (2021).

exception is the (free) learning from other countries' experiences in dealing with a new disease. Yet, in a globalised world where people are free to move, the impact of COVID-19 on a country does not entirely depend on its own actions to prevent it. Lockdown is needed to break the infection chain and also for safeguarding the human health against the virus, but this lockdown has also impacted the economy as a whole; therefore, a strict protocol should be manifested in order to control the spreading (Gauttam et al., 2021).

Given the present scenario the latest National Health Policy (2017) has projected on the "Health in All" where the expenditure in healthcare was increased to a rate of 2.5% of GDP and has expanded its health infrastructure to support the doctors, paramedics, research and development along with hospitals.

REFERENCES

1. Victor, V., Karakunnel, J.J., Loganathan, S. and Meyer, D.F., 2021. From a recession to the COVID-19 pandemic: inflation–unemployment comparison between the UK and India. *Economies*, *9*(2), 73.
2. Danielli, S., Patria, R., Donnelly, P., Ashrafian, H. and Darzi, A., 2021. Economic interventions to ameliorate the impact of COVID-19 on the economy and health: an international comparison. *J Public Health*, *43*(1), 42–46.
3. Gauttam, P., Patel, N., Singh, B., Kaur, J., Chattu, V.K. and Jakovljevic, M., 2021. Public health policy of India and COVID-19: diagnosis and prognosis of the combating response. *Sustainability*, *13*(6), 3415.
4. Fuentes, R., Galeotti, M., Lanza, A. and Manzano, B., 2020. COVID-19 and climate change: a tale of two global problems. *Sustainability*, *12*(20), 8560.
5. National Health Policy. (2017). Ministry of Health and Family Welfare.

11 Disaster Scepticism

CLIMATE CHANGE PATTERN

Arctic ice is melting faster than predicted by the fourth report of the Intergovernmental Panel on Climate Change (IPCC, 2007a). This has led to an unprecedented change in the climate. This change is further elucidated by global or a regional climate pattern which evolves with the increase and decrease in temperature in severe weather events. The change will pave the way for more intensity and frequency of disasters, further giving a way to mega disasters. To address the change of climate and mitigating its impact is the most integral part of sustainable development apart from building a bridge to recede the gap between disaster management, sustainable development, climate change and adaptation. Resilience is a concept that is used in disaster management as a metaphor to define the responses of those affected and those that have responded (Manyena, 2006).

According to the World Health Organization (WHO), the death caused by COVID-19 has surpassed around 18 times more than the havoc of epidemic created during 1970 and 2019 in Latin America and Caribbean by cholera and dengue. Climate change has been a more dangerous threat than the recent pandemic. As per the study of the United Nations (UN) and WHO, it was mentioned that pandemic resulted from the destruction of nature and its biodiversity. The coronavirus pandemic is the repercussion of human intervention and depletion of our natural resources and ecosystem. The illegal wildlife trade and devastation of the forests have hampered the natural habitat of the animals and this has insurged diseases from wildlife to humans.

The biggest problem is coming soon in the form of climate change. The uncontrolled pollution, exploding population and the need for global industrialisation have brought the present on the brink of destruction. The current situation is focused on COVID-19 pandemic which put life in a standstill mode, but the bigger problem is brimming for the future in the form of climate change.

The pandemic has an intertwined crisis and there are climate-induced disasters that have created havoc in the developing nations. There was a series of cyclones that happened in India from early June 2020 to early June 2021. Cyclone Nisarga

DOI: 10.1201/9781003120629-11

struck a few weeks later after Cyclone Amphan which perished more than 80 people in eastern India and Bangladesh.

The rising temperature and locust swarms have devastated the crops and have threatened the security of food and livelihood.

The Red Cross warned East Africa in May 2020 about natural calamities in the form of flood and locust swarms amidst pandemic. The temporary decline in pollution levels during the lockdown gave us the vision that our need for economic growth and overuse of resources has intertwined the environmental crisis.

Increasing populations in urban areas pose high-risk factors such as storm-exposed coasts, flooded deltas of rivers, valleys becoming prone to earthquakes and an increase in volcanic slopes. The cities do not have proper drainage and flood protection systems; instead, they are dependent on essential water and energy supplies.

ENVIRONMENTAL DISRUPTIONS AND LIMITATIONS

Environmental buffers and natural resources overuse have caused huge climatic disturbances apart from these the current pandemic situations along with the challenges faced by governments for large-scale evacuation and temporary housing measures after concurrent natural disaster events has aggravated the situation.

During the pandemic it is difficult to provide people enough shelters for long-term housing and safeguard them from the tropical cyclones or inland flooding. The destructive behaviour we have towards nature has risked our own health. This is the stark reality that we have been avoiding for the past few decades. As per the reports of World Wide Fund (WWF) the major reasons for the diseases to move from the wildlife to humans are majorly due to disruption of the natural habitat, intensive agriculture, huge livestock production and trading along with consumption of high-risk wildlife. The emergence of deadly and unknown variants of diseases have increased over the years due to rapid thawing of ice.

Stringent laws should be augmented by the government to introduce and enforce laws to eliminate the destruction of nature from supply chains of goods and implement sustainable forms of life.

The disaster management depends on three pillars:

1. Prevention – identify the threat;
2. Planning – reduce exposure;
3. Responding – apply lessons learnt from past events to significantly improve the recovery.

However, we are living in an era where man-made disasters have created more havoc than natural calamities. The world has survived the wrath of cholera and dengue but the present pandemic (COVID-19) has impacted the whole world in a very dangerous way.

There is a need to create a strong resilient foundation against biological hazards and pandemics and this year (2021) has taught us the lesson for strengthening sustainability and dealing with global actions to accelerate the transformation required for achieving the 2030 agenda.

HEALTHCARE MANAGEMENT

The utilisation of Health Emergency Disaster Risk Management (Health EDRM) framework has catered to the present pandemic and also solicited to admonish the future risk of this proportion.

• The sustainable development (SDG) goal specifies "good health and well-being" that clearly emphasises early warnings, risk reductions and management of national and global health risks (WHO, 2015). The Paris Agreement along with the IPCC (2007b) report has further elucidated the health risk due to climate change.

The Health EDRM proposes disaster management through the combination of:

a. Hazard and vulnerability reduction to prevent and mitigate risks;
b. Preparedness;
c. Response;
d. Recovery measures.

According to Djalante et al. (2020), the measure taken to strengthen the response to COVID-19 is done by the implementation of Sendai Framework for Disaster Risk Reduction (SFDRR), where knowledge is generated from the Health EDRM framework.

Strategies and mechanisms to cope with disaster resilience give the chance to improvise the probable risks, making a firm foundation for risk governance and enhancing community-based activities.

SOLUTIONS FOR DISASTER MANAGEMENT

There are few suggestive solutions to cope with disasters pertaining to climate change and one such solution is conserving the natural resources to provide nature-based solutions, i.e. preserving the forest cover, coral reefs and wetlands and preparing the communities to cope with and recover from disasters such as drought.

Saving the forest cover and vegetation will help in stabilising the slopes and curtailing the landslide. Wetlands can help in flood regulation. Protection from storm surges, strong winds and cyclones can be provided through the implementation of coastal vegetation and natural biodiversity like mangrove and sand dunes which save from strong storms and cyclones. The thriving coral reef can help in the reduction of wave energy formed during coastal storms. These solutions based on implementation of nature generate local employment and economic opportunities, reducing the need for importing the technical expertise and labour-intensive construction, and therefore offer green solutions.

Preparation for risk reduction and proactive investment will sustain the countries to strive through the disaster and prepare them for the invincible crisis. The inability to counter the risk will falter the secured long-term investments in climate change mitigation and adaptation efforts. The ecosystem management will pay for the longer-term resilience to climate change.

Human activities play a major role in saving our ecosystems. The insatiable greed for more economic gains has paved the path for rapid deforestation. In a study conducted by the European Environment Agency it was found that forests help in soaking up the excess rainwater and prevent run-offs and damage. The presence of a tree prevents the land from excess soil erosion and accumulation of water, thereby preventing landscape flooding. The use of satellite technology in many European nations helped the countries to monitor the rate of deforestation and combat the detrimental effects of climate change, mitigate natural disaster and help in disaster management.

To manage the disaster risk and cope with future calamities with the experience gained in the past led to the formation of disaster management. Climate change in recent times has added more uncertain situations for the assessment of hazards and vulnerability (Figure 11.1). The uncertain present has added greatly to the assessment of hazards and vulnerability. Climate-related risk management occurs at various timescales and can be learnt through the assessment of responses, interventions and recovery that were augmented in earlier impacts. Disaster risk

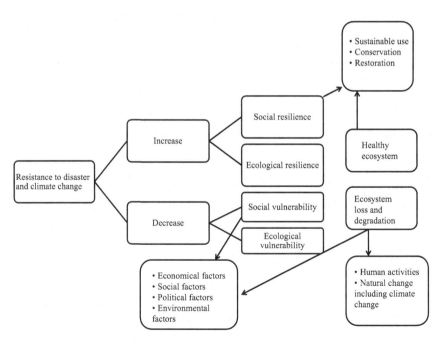

FIGURE 11.1   Ecosystem resilience towards disasters and climate change.

management can be well understood through improvising the efforts administered in various strategies that were implemented to overcome the situation. A concrete plan that encompasses envisioning beyond the current situation will help to deal with the environmental disaster and rapid climate change faced in the present era.

REFERENCES

1. IPCC., 2007a. *IPCC fourth assessment report, working group I report the physical science basis*. Intergovernmental Panel on Climate Change.
2. IPCC., 2007b. *IPCC fourth assessment report, working group II report impacts, adaptation and vulnerability*. Intergovernmental Panel on Climate Change.
3. Manyena, S.B., 2006. The concept of resilience revisited. *Disasters*, 30(4), 434–450.
4. Djalante, R., Shaw, R. and DeWit, A., 2020. Building resilience against biological hazards and pandemics: COVID-19 and its implications for the Sendai Framework. *Progress in Disaster Science, 6*, 100080.
5. www.iucn.org/resources/issues-briefs/nature-based-solutions-disasters.
6. www.eea.europa.eu/highlights/forests-can-help-prevent-floods.
7. World Health Organization, 2015. *Reducing global health risks through mitigation of short-lived climate pollutants. Scoping report for policy-makers*. Geneva: WHO.

12 Healthcare Haven

TRANSVERSE THE CHANGE AMIDST COVID-19

The nexus between COVID-19 and climate change is unravelled through this chapter as climate change according to sustainable development goal 13 that poses a crucial and serious matter as it brought catastrophic incidents bestowing irreversible effects. The pandemic has been an eye-opener for the future of our healthcare as it has raised various issues pertaining to climate change and the environment.

Climate change affects all the three epidemiological triad – humans (the host), pathogen (the agent) and their vectors (the environment) by climatic change like temperature, humidity, and food and water availability.

The effective management of healthcare for the unpredictable climate risks depends on the availability of reliable climate information and services. Co-developing climatic services for a resilient public health system helps a nation to augment the adaptation measures to be taken to safeguard against unprecedented health hazards.

Goal 3 of the sustainable goals of 2030 is focussed on "Good Health and Well-being" and goal 13 on "Climate action" needs to be given a little more emphasis in present times as the world is facing the pandemic along with natural calamities in many continents.

In a study conducted by Barouki et al. (2021) it was found that the development and spread of COVID-19 appeared because of random development in urban areas, depletion of habitat, increasing environmental change, illegal trafficking of animals and a significant rise in global travel.

According to Perkins et al. (2021), there are six important lessons that can be drawn from this pandemic:

1. Comprised the reduction in the usage of fossil fuel and greenhouse emissions
2. Rapid response
3. Sustainable development
4. Trust in science

DOI: 10.1201/9781003120629-12

5. Limits of rugged individualism
6. Opportunity for greater change.

In a study conducted by Fernández-Ahúja and Martínez (2021) a strong inter-connection was found between COVID-19 and environmental elements. The colder regions suffered more compared to warm regions. Environmental factors such as temperature, air pressure, rainfall and sunlight hour have a strong link with the spreading of pandemic.

Pearson and Foxon (2012) noticed that due to industrialisation there is a rapid environmental change that has occurred as the requirement for lower priced products has increased. The change for development had become the root cause for all the possible premature death as massive industrial growth has polluted the air, water, and land through industrial waste deposition.

Poor air quality increases the risk of mortality from severe and chronic diseases such as stroke, heart disease, chronic obstructive pulmonary disease, lung cancer and acute respiratory infection.

MORTAL CONSEQUENCES INTERRELATION

According to Bornstein (2020), environmental pollution is the major reason for mortality compared to malaria, HIV/AIDS, drugs, alcohol, and traffic accidents.

Roser et al. (2020) gave the information about the death of people due to COVID-19 outbreak.

According to Wu and Nethery (2020), 78% of deaths in Italy, France, Spain and Germany are caused in the five most polluted areas of these nations. Factors that affect climate change will be from natural and anthropogenic factors, ocean mesoscale, radiative forces and CO_2 emissions from greenhouse gases. Ocean submesoscale is defined as a region in the ocean where change in global heat leads to climate change (Su et al., 2018). The submesoscale is crucial for the transport of heat between the interior (ocean) and the atmosphere, and thus plays a crucial role in determining the Earth's climate.

STEPS FOR REFORMATION OF ENVIRONMENT

The regional integrated multi-hazard early warning system (RIMES) has been de-vised for Asia and Africa, which is an international and intergovernmental insti-tution for the generation and application of early information. Including India, RIMES has 12 member nations from Asia and Africa along with 19 collaborating countries.

It covers six thematic areas:

•   Weather and extreme events
•   Water-related hazards
•   Seasonal climate

- Earthquake and tsunamis
- Climate change
- Social application.

It provides key services to member states, their New Member States (NMS) and national and local level institutions across thematic areas.

Ten Commandments for co-developing climate services for resilient public health system are the following:

1. Identifying common grounds
2. Cost–benefit analysis
3. Convening convergence
4. Enabling institutions
5. System approach
6. Evidence-based decision-making
7. Interdisciplinary research
8. Pandemic preparation
9. Capacity building
10. Understanding the complexity of adaptation.

For implementation of adaptation measures, synergy between climate change and the health sector is required. To identify the possible gaps and accordingly the suggestion for co-developing the roadmap in climate services and public health sector needs to augmented.

The radiative forces emitted by anthropogenic activities consist of greenhouse gases such as CO_2, methane, N_2O and chlorofluorocarbons along with tropospheric ozone as discussed by Stevenson et al. (1998).

The depletion in the ozone layer is due to stratospheric ozone forces (Andronova and Schlesinger, 2000). The injection of SO_2 gases into the stratosphere is through major volcanic eruptions, therefore creating sulphate aerosols from the stratospheric layer which scatter the incident solar radiation back to space (Andronova and Schlesinger, 2000).

THE PANDEMIC AND ITS ENVIRONMENTAL REPERCUSSIONS

The COVID-19 pandemic has a relation with climate change as significant changes were found in climate due to the pandemic. Some studies suggest that climate change could have led to the emergence of the novel coronavirus (Ching and Kajino, 2020). Nitrogen (NO_2) is a dangerous gas released in air by burning of gasoline, coal, diesel in vehicles, power plants and industrial facilities. According to the latest studies by NASA (2020), NO_2 present near the ground can turn into ozone apart from making air hazy and unsuitable for breathing. The inability to breathe properly causes lung cancer in the long run (Al-Ahmadi and Al-Zahrani,

2013). The lockdown imposed around the world has been a boon for the environment as there was a rapid decrease in nitrogen emission.

A report by Friedlingstein et al. (2020) showed a decline of 2.4 $GtCO_2$, and new estimates suggest that emissions will be regulated at 34 billion tons of CO_2GtCO_2; this has been the largest ever drop recorded according to researchers. Over the years, the magnitude and spatial distribution of geographical distribution of infectious diseases in some plants and animals have altered due to climatic change (Havell et al., 2002).

Though in the present unprecedented time we have witnessed many cases that directly link disease epidemics with climate change, the cause–effect relationship between the both entails an interplay of genetic, biological, demographic, socio-economic, political and technological determinants.

REFERENCES

1. Barouki, R., Kogevinas, M., Audouze, K., Belesova, K., Bergman, A., Birnbaum, L. et al., 2021. The COVID-19 pandemic and global environmental change: emerging research needs. *Environ Int, 146* 106272.
2. Perkins, K.M., Munguia, N., Ellenbecker, M., Moure-Eraso, R. and Velazquez, L., 2021. COVID-19 pandemic lessons to facilitate future engagement in the global climate crisis. *J Clean Prod, 290*, 125178.
3. Pearson, P.J. and Foxon, T.J., 2012. A low carbon industrial revolution? Insights and challenges from past technological and economic transformations. *Energy Policy, 50*, 117–127.
4. Bornstein, M.H., 2020. *Psychological insights for understanding COVID-19 and families, parents, and children.* London: Routledge.
5. Roser, M., Ritchie, H., Ortiz-Ospina, E. and Hasell, J., 2020. Coronavirus pandemic (COVID-19). *Our world in data.*
6. Wu, X., Nethery, R.C., Sabath, B.M., Braun, D. and Dominici, F., 2020. Exposure to air pollution and COVID-19 mortality in the United States. *MedRxiv.*
7. Stevenson, D., Johnson, C., Collins, W., Derwent, R., Shine, K. and Edwards, J., 1998. Evolution of tropospheric ozone radiative forcing. *Geophys Res Lett, 25*(20), 3819–3822.
8. Andronova, N.G. and Schlesinger, M.E., 2000. Causes of global temperature changes during the 19th and 20th centuries. *Geophys Res Lett, 27*(14), 2137–2140.
9. Ching, J. and Kajino, M., 2020. Rethinking air quality and climate change after COVID-19. *Int J Environ Res Public Health, 17*(14), 5167.
10. Al-Ahmadi, K. and Al-Zahrani, A., 2013. Spatial autocorrelation of cancer incidence in Saudi Arabia. *Int J Environ Res Public Health, 10*(12), 7207–7228.
11. Friedlingstein, P., O'Sullivan, M., Jones, M.W., Andrew, R.M., Hauck, J., Olsen, A. et al., 2020. Global carbon budget 2020. *Earth Syst Sci Data, 12*(4), 3269–3340.
12. NASA., 2020. *Nitrogen dioxide levels rebound in China.* Washington, DC: NASA.
13. Harvell, C.D., Mitchell, C.E., Ward, J.R., Altizer, S., Dobson, A.P., Ostfeld, et al., 2002. Climate warming and disease risks for terrestrial and marine biota. *Science, 296*(5576), 2158–2162.

14. Fernández-Ahúja, J.M.L. and Martínez, J.L.F., 2021. Effects of climate variables on the COVID-19 outbreak in Spain. Int J Hyg Environ Health, 234, 113723.

15. Su, Z., Wang, J., Klein, P., Thompson, A.F. and Menemenlis, D., 2018. Ocean submesoscales as a key component of the global heat budget. Nature *Commun*, 9(1), 1–8.

13 Metamorphosis

INNOVATION AND INVOCATION

These two facts (innovation and invocation) refer to invoking the innovator within and creating sustainable options to mitigate the harmful impact of climate on the environment.

Due to lockdown many countries had to face the consequences on their people and economies. Global environmental changes such as soil degradation, ozone layer depletion, pollution and urbanisation have created an inevitable threat to mankind and Earth (Chakraborty and Maity, 2020).

A healthy environment relies on natural life which includes clean air, water and food. Deforestation, intensive and polluting agricultural practices, unsafe management and widespread hunting of wildlife have further aggravated the consequences (Ghebreyesus, 2020). After the pandemic when the situation will be normal and people will be returning back to their usual daily routine, there will be difficult situations for the environment to adapt to the fresh bouts of increasing pollution.

The global problem of climate change and the pandemic are the two concerns that become the cynosure in present times. Presently the prime concern pertaining to the environment lies in the risk of infection. There are various mitigation factors that will be important in the upcoming years to reduce the prevalent concerns.

MITIGATION ROUTE

We reduce the demand on nature by optimising the consumption and making ourselves sustainable by strengthening our local supply chain.

Creating alternatives for animal protein with plant-based proteins and reducing the pollution is essential. The upliftment in the digital revolution was a boost during the pandemic as this will reduce the transportation cost and hence emissions in many sectors.

The global temperature if not controlled now will increase by more than 1.5°C and cause further damage to the environment as a whole. Various sustainable technologies to eradicate and control the climatic changes were implemented. Some of

DOI: 10.1201/9781003120629-13

them are further elucidated – the wetlands that cater to the wastewater and sludge management, the use of organic fertiliser and few more examples.

WETLANDS

Wetlands are the area with a water table near the surface of the land and they sustain seasonally or permanently throughout the year. The global existence of wetland is found in almost all countries (except Antarctica) and they can thrive in any type of climatic conditions.

Wetlands consist of 20–30% of the world's carbon pool and cover about 5–8% of the world's land surface (Mitsch et al., 2013). Apart from other terrestrial ecosystems, wetlands have the highest density that plays an important role in global biogeochemical and carbon cycle along with climate change (Chmura et al., 2003).

With rapid urbanisation there has been unprecedented change in global climate since the 1950s. The surface of the earth has witnessed warmer climate for three decades successively. Between 1880 and 2012 the land and ocean surface temperatures have increased by approximately 0.85°C (Pachauri et al., 2014).

These wetlands play an important role in climate changes because they augment the capacity to modulate atmospheric concentrations of greenhouse gases (GHGs) such as methane,CO_2 and NO_2, which are dominant GHGs contributing about 60%, 20% and 6% of the global warm potential (IPCC, 2007). Wetlands play the role of an important environmental resource as they act as an atmospheric carbon sink that stabilises the climate. They purify the wastewater in a natural way and therefore reduce the use of conventional methods of treatment which uses chemicals and consumes energy for manifesting the treatment process.

The intact wetlands act as buffers in the hydrological cycle (Bergkamp and Orlando, 1999; Junk et al., 2013) and they liberate organic carbon from the atmosphere, neutralising the effects of atmospheric increase in CO_2 (Junk et al., 2013). Wetlands are further defined as constructed wetlands (CWs). CW is considered as a new technological development (Stefanakis, 2019). The conventional wastewater treatment units are energy intensive and costly to build and operate. The nature-based solution used acts as a decentralised wastewater treatment system that appears nowadays as an appropriate alteration for small and medium agglomerations. CW is a cost-effective solution with the lower overall cost for the operation phase of these treatment facilities (Stefanakis et al., 2014; Ghrabi et al., 2011; Gunes et al., 2011). The operational costs could be even up to 90% lower compared to a conventional treatment technology. The warm climate also favours plant metabolism and growth in the wetland system. CW also contributes to sludge management and handling, particularly in remote locations. In conventional system the sludge removed were further processed by sludge drying beds and the end product doesn't have any further application or reuse, whereas the wetland sludge provide an effective dewatering and management solution for minimising day-to-day labour needed to remove the sludge so that the sludge can be effectively used for agricultural purpose (Stefanakis et al., 2011, 2014, 2018; Kengne et al., 2011).

Climate change and its impacts may affect the environment in diverse ways. It causes many unchangeable effects that manifest in species for adapting, migrating or perishing, thereby leading to ultimate loss in local population (Boer, 2012).

A sustainable environment aims at reducing environmental degradation through the process of alteration and reduction of anthropogenic activities. According to Mezher (2011), global warming is considered to be one of the most stressful issues that humanity is facing currently.

AGRICULTURE

Agriculture plays a very significant role in climatic change as the generation of toxic residues through emission, run-off, etc., occurs in this sector. The need for good yield has always been the prime concern considering the humongous population of some countries; therefore, there is widespread use of chemical fertilisers, conventional tillage methods, pest resistant crops and genetically modified seeds.

Organic fertilisers help to mitigate climate changes as they reduce the global emissions from crop and livestock agriculture. Organic fertilisers are used for agriculture purposes and they help to sort out climate change by reducing emissions. There is a direct correlation between nitrous oxide emissions and the amount of nitrogen fertiliser applied to agricultural land. It causes harmful impact on the climatic conditions when synthetic fertilisers are used. Organic farming does not use harmful fertilisers and focuses on creating closed nutrient cycles and allows minimising losses via run-off and emissions because nitrogen generated in organic farms is comparatively less than that in the conventional farms.

The energy use is significantly lower in organic farms as conventional agriculture uses vast quantities of synthetic fertilisers and pesticides. These require loads of energy to manufacture, whereas organic agriculture minimises energy consumption by 30–70% per unit of land. The less energy requirement will also curtail the fuel consumption for transport in organic farming.

The pest-resistant crops are also laden with spurious chemicals and the natural diversity of the crop is compromised, which invariably affect the soil and pose challenges. The soil becomes resilient to floods, droughts and land degradation. According to the Food and Agricultural Organization (FAO) of the United Nations (UN), the overuse of synthetic pesticides and fertilisers needs to be stopped for the betterment of the environment.

Integrated farming should be promulgated where farmers instead of purchasing fertilisers can switch to the natural production of the required organic fertiliser as defined in the example given below.

Increase in human population has made us confront so many challenges, such as food shortages, malnutrition, limited means of irrigation and shrinking land resources, along with deteriorating environmental quality. To cater to the increasing demand of food the intensification of the agricultural production system is needed. Around 50% of the population depends on rice and it has always been the major staple food of people in Asia.

The production office requires a huge amount of water resources and the paddy fields emit large amounts of GHGs, i.e. methane. Rice farming needs to upgrade the management and incorporation of rice–fish farming will give the opportunity to face such challenges. Rice cultivation impacts the environment as it generates huge amounts of GHGs which consist of methane and nitrous oxide.

Methane is produced by anaerobic degradation of organic complexes such as plant residues, organic matter and organic fertilisers under submerged conditions where oxygen lacks. About 10–20% of methane in the atmosphere is generated from paddy fields. The global warming potential (GWP) of methane is 25 times more compared to carbon dioxide (CO_2).

When rice is cultivated along with fish, it lowers the emission of methane and other GHGs. Aquatic animals like carp fish and crabs disturb the soil layers by their continuous movement in search of food and therefore influence the methane production.

Aquatic animals increase the diluted oxygen in the field water and also in the soil, thereby shifting the anaerobic digestion to aerobic digestion and thus decreasing the emission of methane. Around 34.6% less methane was generated due to this innovative and frugal method of co-cultivation. This mode of cultivation has increased the revenue generation, and there is an increase in the yield by 25% and less labour cost (i.e. around 19–22%), with a 7% lower input for material for the farmers. This co-culture cultivation will reduce the environmental degradation and will give economic stability to the farmer (Ly et al., 2020).

URBANISATION

With this type of environmentally sustainable cultivation it also promulgates the need for an economically feasible and environment-friendly dwelling.

Anthropogenic growth has been prevalent with the increasing need for land resources to create concrete jungles. Rapid urbanisation has deforested many jungles and invariably harmed the environment. The need for a sustainable dwelling is the need of the hour.

According to the National Aeronautics and Space Administration (NASA) and National Oceanic and Atmospheric Administration, 2020 is recorded as the hottest year. As per the international scientific community human activity is the key driver for global warming. As per the Intergovernmental Panel on Climate Change (IPCC), human activity is 95% responsible for global warming for the past 50 years. Buildings account for 40% of all GHG emissions as per the reports of 2030 Architecture (Barnes and Parrish, 2016). Other activities such as transport are also associated with buildings.

By incorporating green buildings we can curtail the harmful effect of buildings on climate change. Green building with its structural design, operation, construction and cradle to grave concept for materials used will create a benchmark for the sustainability along with the location selected for implementing the concept. Green buildings and the communities help in reduction of landfill waste and enable attentive transportation of vegetated land areas. The need for sustainable building caters to the humongous population of Asia and Africa. These buildings consume around 40–50% less energy and save 20–30% water consumption compared to conventional buildings.

As per the UN report for habitat, around 62% fewer GHGs were emitted by low carbon buildings of Australia. Green buildings are generally made out of natural and renewable components like limestone, calcined clay cement, construction and demolition waste, fly-ash, cement-free concrete and even the waste generated from metal industry which are turned into blocks. Apart from these materials made from renewable biowaste and reinforced earth are also being used for this purpose. The lowering of CO_2 emission lowers the production energy, cost-effectiveness and zero GHG emission. Even with advancement in technological era, sometimes it is difficult to predict the result of climate change.

In October 2018, the UN report from the IPCC summarised that the need of the hour is the large, immediate and unprecedented global efforts to mitigate the requirement of GHGs. In the Paris Agreement, 200 nations were bonafide members to curtail the temperature rise to a maximum level of 2°C but IPCC reported that if the temperature rises beyond 1.5°C (2.7F), then there will be a huge catastrophic effect pertaining to the recent situation. Though it may be witnessed that there is a rise in a couple of degrees, the detrimental impact through stronger storms, acidified oceans (loss of coral reef and shellfish) and extreme cold weather will lead to a catastrophic loss of species and ecosystems. The increase in drought and higher temperature will lead to an increase in species migration, thereby leading to hostile situations for survival in a limited available resource.

Our every little action causes a huge impact and this is an unprecedented opportunity to work together. We drastically need to reduce carbon emission and prepare ourselves for climate change.

REFERENCES

1. Chakraborty, I. and Maity, P., 2020. COVID-19 outbreak: migration, effects on society, global environment and prevention. *Sci Total Environ*, *728*, 138882
2. Ghebreyesus, T.A., 2020. Safeguard research in the time of COVID-19. *Nature Med*, *26*, 443.
3. Stefanakis, A.I., 2015. Constructed wetlands: description and benefits of an eco-tech water treatment system. In McKeown, A.E. and Bugyi, G. (eds.), *Impact of water pollution on human health and environmental sustainability* (1st edn, pp. 281–303). Hershey, Derry Township, PA: Information Science Reference (an imprint of IGI Global),.
4. Stefanakis, A.I., Akratos, C.S. and Tsihrintzis, V.A., 2014. *Vertical flow constructed wetlands: eco-engineering systems for wastewater and sludge treatment* (1st edn.) Amsterdam: Elsevier Publishing.
5. Stefanakis, A.I., 2019. The role of constructed wetlands as green infrastructure for sustainable urban water management. *Sustainability*, *11*, 6981.

6. Ghrabi, A., Bousselmi, L., Masi, F. and Regelsberger, M., 2011. Constructed wetland as a low cost and sustainable solution for wastewater treatment adapted to rural settlements: The Chorfech wastewater treatment pilot plant. *Water Sci. Technol.*, *63*, 3006–3012.

7. Gunes, K., Tuncsiper, B., Masi, F., Ayaz, S., Leszczynska, D., Hecan, N.F., et al., 2011. Construction and maintenance cost analyzing of constructed wetland systems. *Water Pract. Technol.*, *6*(3), 1–2.

8. Stefanakis, A.I., Prigent, S. and Breuer, R., 2018. Integrated produced water management in a desert oilfield using wetland technology and innovative reuse practices. In Stefanakis, A.I. (ed.), *Constructed wetlands for industrial wastewater treatment* (1st edn., pp. 25–42) Chichester: John Wiley & Sons Ltd.,.

9. Stefanakis, A.I., Komilis, D. and Tsihrintzis, V.A., 2011. Stability and maturity of thickened wastewater sludge treated in pilot-scale sludge treatment wetlands. *Water Res, 45*, 6441–6452.

10. Kengne, I.M., Kengne, E.S., Akoa, A., Bemmo, N., Dodane, P.H. and Kone, D., 2011. Vertical-flow constructed wetlands as an emerging solution for faecal sludge dewatering in developing countries. *J Water Sanit Hyg Dev*, *1*, 13–19.

11. Boer, H., 2012. Policy options for, and constraints on, effective adaptation for rivers and wetlands in northeast Queensland. *Australas J Environ Manag*, *17*(3), 154–164.

12. Mezher, T., 2011. Building future sustainable cities: the need for a new mindset. Construc Innovat, *11*(2), 136–141.

13. Barnes, E. and Parrish, K., 2016. Small buildings, big impacts: the role of small commercial building energy efficiency case studies in 2030 districts. *Sustainable Cities and Society*, *27*, 210–221.

14. Mitsch, W.J., Bernal, B., Nahlik, A.M., Mander, Ü., Zhang, L., Anderson, et al., 2013. Wetlands, carbon, and climate change. *Landscape Ecology*, *28*(4), 583–597.

15. Chmura, G.L., Anisfeld, S.C., Cahoon, D.R. and Lynch, J.C., 2003. Global carbon sequestration in tidal, saline wetland soils. Global Biogeochem Cycle, 17, 1111.

16. Pachauri, R.K., Allen, M.R., Barros, V.R., Broome, J., Cramer, W., Christ, R., et al., 2014. Climate change 2014: synthesis report. Contribution of working groups I, II and III to the fifth assessment report of the Intergovernmental Panel on Climate Change (p. 151). Geneva: IPCC.

17. IPCC, 2007. *Climate* change *2007: the physical science basis.* Cambridge: Cambridge University Press.

18. Bergkamp, G. and Orlando, B., 1999, October. Wetlands and climate change. In Exploring collaboration between the convention on wetlands and the United Nations framework convention on climate change. Ramsar: World Conservation Unión (IUCN).

19. Junk, W.J., An, S., Finlayson, C.M., Gopal, B., Květ, J., Mitchell, S.A., et al., 2013. Current state of knowledge regarding the world's wetlands and their future under global climate change: a synthesis. *Aqua Sci*, *75*(1), 151–167.

20. Lv, W., Yuan, Q., Lv, W. and Zhou, W., 2020. Effects of introducing eels on the yields and availability of fertilizer nitrogen in an integrated rice–crayfish system. *Sci Rep*, *10*(1), 1–8.

14 Comparative Correlation Through Case Studies

DORMANT HOTSPOTS AWAKENED BY ANTHROPOGENIC ACTIVITIES

The prevalence of some diseases is persistent in developed countries and with human interventions it has increased with time. According to the World Health Organization (WHO) consecutive reports (WHO, 2017, 2018, 2019), the control of malaria in South America is an obstacle that threatens global strategies for its elimination by 2030.

Since the collapse of Venezuelan economy in 2014, the entire healthcare system has been in surge (Grillet et al., 2018, 2019). The death rate reported in Venezuela in 2015–2018 was nearly a total of 1,255,299 cases and the year 2017 exhibited the most substantial increase in malaria cases worldwide (Battle et al., 2019). Conn et al. (2018) in their study found that in the endemic countries of South America some of the species of malarial infection are prevalent, i.e. *Plasmodium vivax* which accounts for 76%, *Plasmodium falciparum* (17.7%), mixed *P. vivax/ P. falciparum* infections (6%) and *Plasmodium malariae* (< 1%) (WHO, 2019). Gabaldon (1983) has mentioned the eradication in the early 1960s. Magris et al. (2007) reported that some of the strains like *P. falciparum* and *P. vivax* are prevalent in the lowland of Amazon rainforests and Savannas of the remote Guayana region. *P. vivax* malaria re-emerged along the coastal wetlands and the transmission was interrupted 20 years later (Grillet et al., 2014). According to Moreno et al. (2014) and Barrera et al. (1999), some regions of southeastern Venezuela like Bolivar state have contributed >60% (1992–1995) to 88% (2000–2014) of total cases of Malaria. The clearing of forest due to gold mining has accounted for around 47–80% of malarial cases (Moreno et al., 2014; Recht et al., 2017). The workers of gold mines augmented for 47–80% of malaria cases (Recht et al., 2017; Conn et al., 2018; Moreno et al., 2014). The local transmission of malaria, which was endemic, has re-emerged since 2014 (Grillet et al., 2019).

The connecting corridor of Bolivia to other states of Brazil will spread the disease, giving rise to the malaria epidemic (Pacheco et al., 2019). The situation is aggravated when there is an artemisinin resistance (Chenet et al., 2016) observed due

DOI: 10.1201/9781003120629-14

to the presence of novel mutations which lead to the delayed parasite clearance in around Venezuela and transforming the area into a hub (Pacheco et al. 2020). The gold mining has promulgated the rise of malarial infections along with sustaining and restoring the transmission. Human intervention has given rise to so many health gaps and they have invariably caused a lot of damage to the ecosystem/ habitat. In the Fourth Assessment Report of the IPCC it was reported that limited parameterization was the key factor which has influenced the geographical range and intensity of the malarial transmission. The situation has further aggravated when the need for gold mines for economic gain is more than the human lives.

BIOMEDICAL WASTE (BMW) GENERATION HAS AGGRAVATED THE PANDEMIC

In the current pandemic situation, the waste generated in the form of solid and liquid is very high. Developed and developing nations have depicted certain methods of disposal of solid waste as defined in Figure 14.1.

For a densely populated country like India the generation of BMW has aggravated the pandemic situation over the last seven months. The production of waste was 33,000 tons and daily generation of 146 tons of COVID waste across the

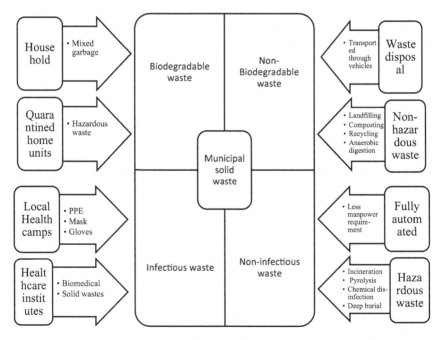

FIGURE 14.1   The flowchart of the different policies adopted by developing and developed countries for processing of municipal solid waste.

nation was alarming for the nation. The spike in the COVID-19 waste production was due to the following reasons:

1. Generation of huge amounts of medical wastes from healthcare centres as a huge population was tested positive. These wastes include cotton swabs, injections, and PPE kits.
2. The majority of wastes generated from households are infectious in nature as self/home quarantined patients are being dumped in the common waste collection area.

Out of the 36 states/union territories (UTs) in India, only eight states have proper healthcare facilities and stringently follow the BMW management rules. And though the deep burial methods are forbidden in India, still 23 states/UTs are using this method (Table 14.1).

This table gives us the estimated use of mass across the globe and the various consequences faced for its disposal.

Total daily facemasks = (#population *#urban population (%) *##facemask acceptance rate (%) *##average daily facemasks per capita)/10,000 (Tripathi et al., 2020).

The disposal of biomedical waste plays an important role in safeguarding the environment. Described herewith are some of the technologies implemented for biomedical waste treatment (Khadem et al., 2016):

1. Landfilling – This is the most primitive method of disposal of biomedical waste in undeveloped nations. These types of disposal cause contamination of soil and water, spreading of pathogens and emission of greenhouse gases (GHGs). Though it is economical and easy to operate, the risk of contamination is high.

TABLE 14.1

The Estimated Quantity of Masks to be Used During a Pandemic (Nzediegwu and Chang, 2020)

Continents	Total daily facemasks (million)	Weight of total daily facemasks used (tons)
Asia	3,716.20	1,486.48
Africa	922.22	368.89
Europe	884.71	353.88
North America	489.05	195.62
Oceania	45.43	18.17
South America	544.39	217.75

2. Sterilization – This method is preferred in many countries as it provides excellent efficiency, less treatment time, low cost, less emission of GHGs apart from being environment friendly.
3. Incineration – The biomedical waste incineration is suitable for all waste types. This method reduces the volume of the waste and hence has high potential for recovery as well as providing safe disposal.
4. Microwaving – This is the most efficient and environment-friendly technology, which reduces the waste volume, does not generate liquid waste and minimizes air pollutants. High cost, selective waste treatment, odour problems and high emission of GHGs are some of the disadvantages.
5. Plasma pyrolysis – This is suitable for all waste types, occupies less space, is environmentally-friendly, chimney is not required for emission, toxic residuals are minimum, doesn't require segregation of wastes, energy recovery, and waste volume reduction by 90%.

The effect of biomedical waste is detrimental and the impact of the same is described in the case study of Pune, India. Pune is one of the largest metropolitan cities of India in the state of Maharashtra and has a population strength of nearly 60, 49,968 as per the 2001 census.

The city augments a huge number of hospitals, i.e. around 928, and caters to nearly 3,350 patients on daily basis as per the corporation of Pune. The total amount of bio-waste generated is around 3,000 kg, with 450 kg generated by corporation-run hospitals. The Pune Municipal Corporation receives biomedical waste from 764 hospitals, 2,222 clinics, 222 pathology laboratories and 12 blood banks and only 2,162 clinics send it to biomedical waste treatment facilities as per the reports of Indian Express news daily. Apart from this, nearly 5,000–7,000 medical practitioners are in Pune (including Homeopathic, Ayurvedic, Unani and other practitioners), and around 1,000 have registered for biomedical waste treatment facilities.

The lack of proper biomedical waste treatment has become a threat to the hospitals and clinics in Pune. Though law prescribes a proper treatment procedure, the lack of knowledge and the importance of biomedical waste treatment often dump the waste in garbage bins. This leads to the spread of communicable diseases and often leads to the emergence of epidemics. As installing an incinerator is a costly affair, this facility is not provided in most of the smaller hospitals and clinics. This paves the way for untreated disposal where the biomedical way of disposal is not followed, which leads to a lot of problems in the long run (Acharya et al., 2014).

With the present situation of the pandemic, this type of practice has aggravated the situation and caused a widespread infection in the city.

Figure 14.2 shows the state-wise distribution of COVID-19 biomedical waste (BMW) in the pandemic situation and the above cited case study of Pune correlates for the burgeoning waste in the present situation and the lack of treatment options has generated a humongous amount of biomedical waste.

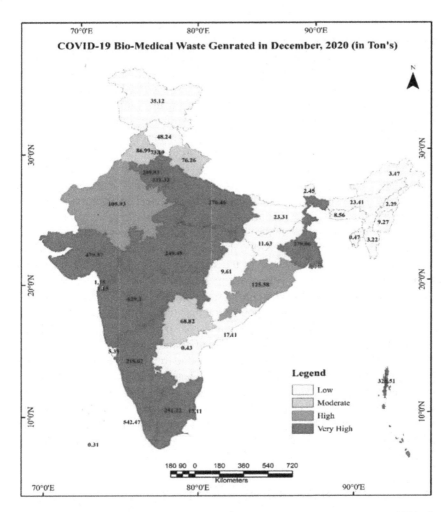

FIGURE 14.2 State-wise distribution of COVID-19 BMW in December 2020 (in metric tons).

Source: Rajak et al. (2021).

THAWING PERMAFROST AND THE CLIMATE RESILIENCE

Permafrost plays an important role in keeping the dangerous environmental factors frozen below the ground. It is mainly found in the Arctic region, 80% in Alaska, 50% in Canada and 60% in Russia. They are also found intact in many ice-laden colder regions of many nations. It has three main characteristics, where the first one augments it as a gigantic freezer storing vast amounts of organic matter, remnants of dead plants, animals, microbes and viruses , the second one is the huge amount of methane released from the methane hydrates also known as 'clathrates'.It is also known as methane "ice" that originates at low temperatures and high pressures in the continental margin marine sediments else they are found beneath the permafrost. The third one is pertaining to the huge amount of carbon being released to the atmosphere which is approximately 1460bn-1600bn tonnes of carbon. The carbon released through these permafrost is a sustained and continuous process that adds carbon to the atmosphere thereby increasing the warming of the earth surface.(Schädel, 2020) The regions of military infrastructure with permafrost cause a direct threat and promulgate future thawing. The thawing of permafrost will induce avalanche incidences in Alps that will cause harm to the locals and tourists (Jorio, 2020). Due to the environmental factors leading to the thawing of permafrost several military bases in Arctic regions have collapsed, roads and houses have become unstable and it was shifted to a safer place as per the Special Report on the Ocean and Cryosphere. Thawing is also an underlying problem for oil and natural gas pipes which might get spilled in the coming years or might lead to accidental hazards in future. The biggest fear is that thawing of the permafrost will release the frozen viruses and will contribute to the spread of infectious agents (Legendre et al., 2017). In some studies permafrost melting acts as the source of pandemic-generating viruses. It was prevalent in Russia, those regions having permafrost had more COVID-19 cases per million population than those without them. Permafrost is categorised into continuous, discontinuous or sporadic after it melts (Zhang et al., 2000; Dobinski, 2011; All about Frozen Ground, 2021).

There were five major influenza pandemics due to the effects of monthly sunspots and it implies that permafrost melting releases influenza viruses that were considered novel at that time. After the Second World War there was a sudden increase in temperature where Spanish flu was found prevalent with its roots in Siberia's hot summers. The hot spells from 1912 to 1914 were constantly at a high and were evident before the rapid spread of Spanish flu in 1918 (Saunders-Hasting and Krewski, 2016; Kilbourne, E.D., 2006).

CANADIAN WILDFIRE DUE TO CLIMATE CHANGE

Lytton is a small village of Canada with 250 inhabitants and around 1,000 indigenous population of Nlaka'pamux people also reside in that area. The lethal heatwaves and soaring temperatures (49.6°C/121.3F) have given way for the forest fires, along with winds at 71 km/h that have spread them further. The variation in night and day temperatures due to blistering heatwaves is a harbinger of climate change in the neighbouring areas. The excessive heat has caused several deaths as people are not accustomed to this much variation in temperature in Canada where the average temperature is generally as low as 20°C. Forest fire has aggravated the situation and the whole town experienced a catastrophe of climate change.

Heat action plans were taken to address the heatwave issues and improve interagency collaboration. When the temperature crosses beyond 47°C, it is termed as a severe heatwave. The wet-bulb temperature is a condition when the human body reacts to a combination of heat and humidity. The excessive humidity also causes heat flushes in the body and makes it unbearable for manual labour and the body cannot cool themselves if the temperature soars more than 36°C. The unanticipated heatwave has forced humanity to implement nationwide heat code criteria where standard operating procedures (SOPs) need to be developed to administer the heatwaves, relief measures and availability of healthcare centres.

Measures are to be taken for urban planning and redefining the building codes for the already existing structures/buildings. Urban planning model should be redefined and greenery should be prioritised along water bodies and incorporation of cool roofs and green buildings. Under the National Disaster Management Act of 2005, the Indian government has categorised heatwave as a natural disaster. The urge to take the heatwave as a detrimental effect in the present scenario will also help us to prepare in advance for such disasters. Millions of sudden deaths were witnessed due to severe heatwaves.

REFERENCES

1. WHO 2017. *World malaria report 2017*. Geneva: World Health Organization. Licence: CC BY-NC-SA 3.0 IGO.
2. WHO 2018. *World* malaria rep*ort 2018*. Geneva: World Health Organization. Licence: CC BY-NC-SA 3.0 IGO.
3. WHO 2019. *World* malaria re*port 2019*. Geneva: World Health Organization. Licence: CC BY-NC-SA 3.0 IGO.
4. Grillet, M.E., Villegas, L., Oletta, J.F., Tami, A. and Conn, J.E., 2018. Malaria in Venezuela requires response. *Science. 359*, 528.
5. Grillet, M.E., Hernández-Villena, J.V., Llewellyn, M.S., Paniz-Mondolfi, A.E., Tami, A., Vincenti-Gonzalez, M.F. et al., 2019. Venezuela's humanitarian crisis, resurgence of vector-borne diseases, and implications for spillover in the region. *Lancet Infect Dis, 19*(5), 149–161.
6. Battle, K.E., Lucas, T.C.D., Nguyen, M., Howes, R.E., Nandi, A.K., Twohig, K.A. et al., 2019. Mapping the global endemicity and clinical burden of *Plasmodium vivax*, 2000–17: a spatial and temporal modelling study. *Lancet, 394*(10195), 332–343.
7. Conn, J.E., Grillet, M.E., Correa, M. and Sallum, M.A.M., 2018. Malaria transmission in South America - present status and prospects for elimination. In: Manguin, S. and Dev, V. (eds.), *Towards malaria elimination—a leap forward* (pp. 281–313). London: IntechOpen.
8. Gabaldon, A., 1983. Malaria eradication in Venezuela: doctrine, practice, achievements after twenty years. *Am J Trop Med Hyg, 32* , 203–211.
9. Magris, M., Rubio-Palis, Y., Menares, C. and Villegas, L., 2007. Vector bionomics and malaria transmission in the Upper Orinoco River, Southern Venezuela. *Mem Inst Oswaldo Cruz, 102*(3), 303–311.
10. Barrera, R., Grillet, M.E., Rangel, Y., Berti, J. and Aché, A., 1999. Temporal and spatial patterns of malaria reinfection in north-eastern Venezuela. *Am J Trop Med Hyg, 61*, 784–790.

11. Grillet, M.E., El Souki, M., Laguna, F., León, J.R., 2014. The periodicity of *Plasmodium vivax* and *Plasmodium falciparum* in Venezuela. *Acta Trop*, *129*, 52–60.

12. Moreno, J.E., Rubio-Palis, Y., Martínez Ángela, R. and Porfirio, A., 2014. Evolución espacial y temporal de la malaria en el municipio Sifontes del estado Bolívar, Venezuela. 1980–2013. *Bol Malariol Salud Ambient*, *54*, 236–249.

13. Recht, J., Siqueira, A.M., Monteiro, W.M., Herrera, S.M., Herrera, S. and Lacerda, M.V.G., 2017. Malaria in Brazil, Colombia, Peru and Venezuela: current challenges in malaria control and elimination. *Malar J*, *16*, 273.

14. Pacheco, M.A., Schneider, K.A., Céspedes, N., Herrera, S., Arévalo-Herrera, M. and Escalante, A.A., 2019. Limited differentiation among *Plasmodium vivax* populations from the northwest and to the south Pacific Coast of Colombia: a malaria corridor? *PLoS Negl Trop Dis*, *13*(3), e0007310.

15. Chenet, S.M., Akinyi Okoth, S., Huber, C.S., Chandrabose, J., Lucchi, N.W., Talundzic, E. et al., 2016. Independent emergence of the *Plasmodium falciparum* Kelch propeller domain mutant allele C580Y in Guyana. *J Infect Dis*, *213*, 1472–1475.

16. Pacheco, M.A., Forero-Peña, D.A., Schneider, K.A., Chavero, M., Gamardo, A., Figuera, L. et al., 2020. Malaria in Venezuela: changes in the complexity of infection reflects the increment in transmission intensity. *Malar J*, *19*, 176.

17. Nzediegwu, C. and Chang, S.X., 2020. Improper solid waste management increases potential for COVID-19 spread in developing countries. *Resour Conserv Recycl*, *161*, 104947.

18. Tripathi, A., Tyagi, V.K., Vivekanand, V., Bose, P. and Suthar, S., 2020. Challenges, opportunities and progress in solid waste management during COVID-19 pandemic. *Case Stud Chem Environ Eng*, *2*, 100060.

19. Khadem Ghasemi, M. and Mohd Yusuff, R., 2016. Advantages and disadvantages of healthcare waste treatment and disposal alternatives: Malaysian scenario. *Pol J Environ Stud*, *25*, 17–25.

20. www.punecorporation.org.

21. www.indianexpress.com.

22. Acharya, A., Gokhale, V.A. and Joshi, D., 2014. Impact of biomedical waste on city environment: case study of Pune, India. *J Appl Chem*, *6*(6), 21–27.

23. Rajak, R., Mahto, R.K., Prasad, J. and Chattopadhyay, A., 2021. Assessment of bio-medical waste before and during the emergency of novel Coronavirus disease pandemic in India: a gap analysis. *Waste Manag Res*, *1*, 0734242X211021473.

24. Schädel, C., 2020. The irreversible emissions of a permafrost 'tipping point'. *World Economic Forum*, 18 Feb. 2020.

25. Jorio, L., 2020. La disparition du permafrost, une menace locale, régionale et mondiale. *Swissinfo*, 4 Nov. 2020.

26. Legendre, M., Bartoli, J., Shmakova, L., Jeudya, S., Labadie, K., Adrait, et al., 2017. Thirty-thousand-year-old distant relative of giant icosahedral DNA viruses with a pandoravirus morphology. *Proc Natl Acad Sci USA*, *111*, 4274–4279.

27. Zhang, T., Heginbottom, J.A., Barry, R.G. and Brown, J., 2000. Further statistics on the distribution of permafrost and ground ice in the Northern Hemisphere. *Polar Geog*, *24*, 126–131.

28. Dobinski, W., 2011. Permafrost. *Earth-Sci Rev*, *108*, 158–169.

29. All about Frozen Ground. Available online: https://nsidc.org/cryosphere/frozengro und/index.html (accessed 24 Feb. 2021).

30. Saunders-Hasting, P. and Krewski, D., 2016. Reviewing the history of pandemic influenza: understanding patterns of emergence and transmission. *Pathogens, 5,* 66.
31. Kilbourne, E.D., 2006. Influenza pandemics of the 20th century. *Emerg Infect Dis, 12,* 9–14.
32. www.bbc.com/news/world-us-canada-57678054.
33. https://timesofindia.indiatimes.com/blogs/toi-edit-page/canadas-wildfires-are-a-warning-to-india-here-is-how-we-must-shore-up-our-cities-against-heatwaves/.

Index

Printed in the United States
by Baker & Taylor Publisher Services